A KEY TO THE

FRESHWATER PLANKTONIC and SEMI-PLANKTONIC ROTIFERA of the BRITISH ISLES

by

ROSALIND M. PONTIN

FRESHWATER BIOLOGICAL ASSOCIATION
SCIENTIFIC PUBLICATION No. 38
1978

PREFACE

The rotifers are some of the most abundant of freshwater animals but, apart from the work of a small number of enthusiasts, they have tended to be neglected by professional freshwater biologists. Among the reasons for this are the large number of species and the complexities of their taxonomy. We are therefore grateful to Mrs Pontin for preparing this key to all the species likely to be found in freshwater plankton in Britain and Ireland.

There is good evidence that planktonic rotifers are significant components of freshwater ecosystems. For example, grazing by rotifers can rapidly reduce populations of planktonic algae in lakes and reservoirs. We hope, therefore, that the publication of this key will stimulate more biologists to take an interest in these fascinating animals. Both Mrs Pontin and the Association would welcome new information on the distribution of rotifer species and comments on problems found when using this key.

The Ferry House
August 1978

E. D. Le Cren
Director.

SBN 900386 33 9
ISSN 0367-1887

CONTENTS

INTRODUCTION

A. SCOPE OF THE KEY

The Rotifera are microscopic animals, most of them less than 1 mm long (1000 μm), many species measure about 250 μm, and some less than 100 μm. They are found in open water of lakes, ponds, pools, canals and slow rivers, and also in mud, detritus, moist sand, damp moss and aquatic vegetation. This key is confined to freshwater species. Occurrence of these species in brackish and saline habitats is noted but species found exclusively in such waters are omitted. The key is also mainly confined to planktonic and semi-planktonic species. It is not easy to draw a clear distinction between planktonic and non-planktonic species. While some species inhabit the open water all the time, others may be planktonic at some time of the year or may divide their time between open water and vegetation. Some genera include only truly planktonic species, others have planktonic, semi-planktonic and non-planktonic members. The aim in this key has been to include those species which may be collected by drawing a fine-mesh plankton net through the open water of lakes, ponds, pools or canals, while omitting those species encountered normally in littoral or floating vegetation or in the benthos. However, as it is often difficult to avoid collecting some littoral species when the net is drawn in to the bank, some of the more common genera living in shore vegetation are included. Non-planktonic members of each genus are also referred to in the keys to species. Habitat preferences are indicated for each species, but many species of Rotifera, in particular the planktonic ones, are widespread and ubiquitous. The keys include the recorded British and Irish species, with the addition of some others which may have been overlooked because of difficulties in identification.

The key is intended for the use of those unfamiliar with the structure and habits of the rotifers, and is based as far as possible on those morphological characters which are most easily observed. The main character which has formerly been used to separate most families is the structure of the jaws. As this character is not easy to observe in many rotifers, it is used as little as possible in the present key. There is no separate key to families; the key gives identification to genera and then to species, which are grouped in families.

Most rotifers encountered are females, reproducing parthenogenetically for most of their season of occurrence. Characters given in the main keys and descriptions of species apply mostly to females, but notes on the males are included. A key to genera of males is given separately (p. 148). Males are sporadic, often seasonal. They may be very numerous during their period of occurrence, but seem to be less frequently encountered by collectors, probably because they are always smaller, and often much smaller, than their females. They are usually short-lived but may survive for several days. The males of some rotifers have not yet been found (e.g. *Squatinella*, some species of *Synchaeta*) and none has been found for any of the bdelloid group (which includes *Rotaria*). Reduction in complexity and absence of many of the characteristics of the females, such as spines and appendages, make for considerable similarity between many species, especially within a genus. This, coupled with insufficient information on the males of many species, makes construction of a key very difficult. Careful observations and drawings of any males found, especially of the less well-known species, could be of great value for future keys.

In the illustrations the magnification of each drawing is indicated by a scale-line bearing a figure which is its length in micrometres (μm).

B. MORPHOLOGY OF THE FEMALE ROTIFER (fig. 1)

There is a *head, body* (or trunk) and *foot*, but the head is not very distinct from the body, into which it and the foot can usually be withdrawn. Some species have no foot. The transparency of most rotifers allows the internal organs to be observed in living animals; they may indeed be amongst the most striking features.

Head

There is an apical area which bears the *corona* or wheel-organ. This is the characteristic ciliary organ of the Rotifera, and is usually used for swimming and collecting food. It consists of a circle of cilia around the head (the *circumapical band*) and a ciliated area around the mouth (the *buccal field*). This basic plan is modified in most rotifers to a pattern characteristic of the family or genus, e.g. in Brachionidae the corona consists of arcs and tufts of fused cilia (cirri) around the head and mouth (fig. 1, CC), while in Asplanchnidae the buccal cilia are very reduced and the corona consists of a single girdle encircling the head (fig. 35). The corona may be extended around lobes as in Collothecidae (fig. 27) or on to appendages (*Synchaeta*, fig. 23). The shape and arrangement of the corona may be important as diagnostic features.

The basal bodies and roots of the coronal cilia are contained in cushions of epidermal tissue which also contain one to several nuclei. These cushions are usually conspicuous, lying below the corona and projecting into the body cavity.

The head encloses the brain (fig. 1, BR), from which a system of nerves, some bearing ganglia, radiates to all parts of the body. The head bears many of the sense organs including sensory bristles and pits, finger-like

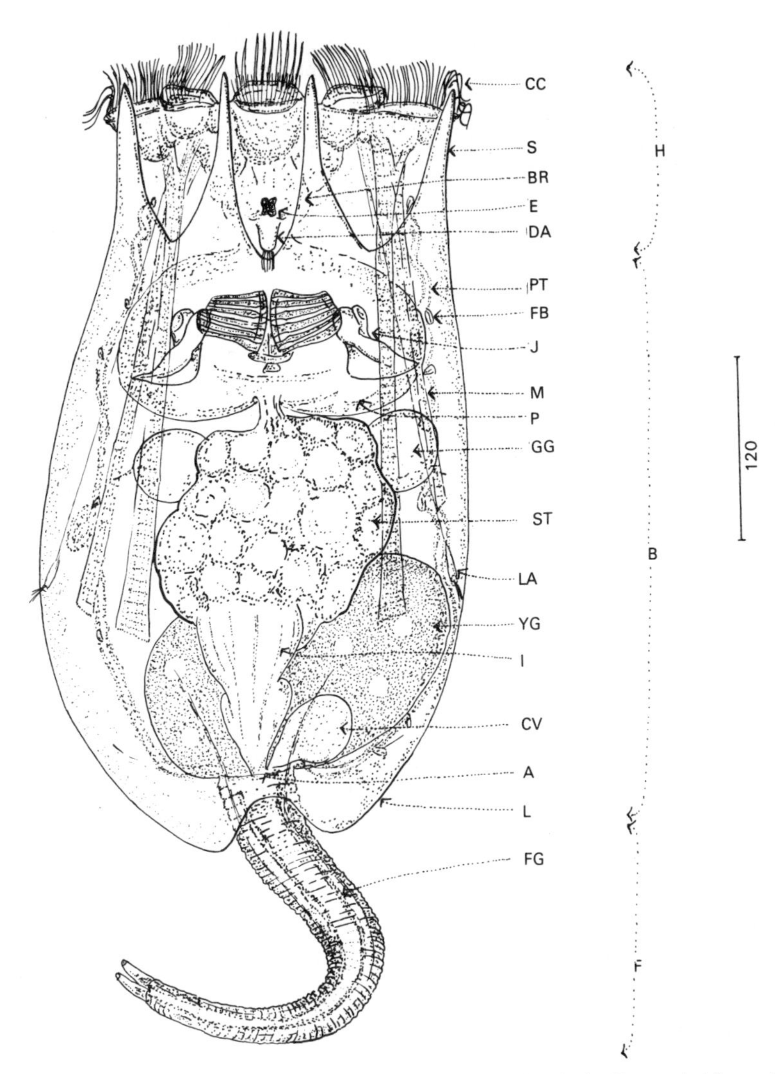

Fig. 1. *Brachionus calyciflorus*. Female, dorsal. H head; B body; F extended foot, with toes; A anus position; BR brain; CC coronal cilia; CV contractile vesicle; DA dorsal antenna; E eye; FB flame bulb; FG foot gland; GG gastric gland; I intestine; J jaws; L lorica; LA lateral antenna; M muscle; P pharynx; PT protonephridium; S spine; ST stomach; YG yolk gland. (Partly after Hollowday, personal communication).

palps and eyes or pigmented areas; there is typically a sensory *dorsal antenna* (fig. 1, DA), which consists of a small bristle-bearing projection, and a pair of similar *lateral antennae* is found on the sides of the body (fig. 1, LA). One, two or three eyes may be connected to the brain by nerves or be situated directly upon it (fig. 1, E). A so-called retrocerebral organ of uncertain function may be present, lying close behind the brain, and may consist of a median sac and paired glands opening on the apical field; it is particularly well-developed in some genera, e.g. *Trichocerca* (fig. 10), and reduced or absent in others.

Body

The body may be cylindrical, sack- or bell-shaped, or laterally or dorsoventrally flattened to various degrees. Its flexibility is dependent on the cuticle or skin, which may be very thin, very flexible and very transparent, or thickened over parts or all of the body. The thickened cuticle is stiffer, though still flexible to some extent, and fairly transparent. In many species the thick cuticle forms a distinct shell or *lorica* (fig. 1, L), of one or more plates, enclosing the body and in many cases the head and foot when these are retracted. The lorica may bear various kinds of 'ornamentation' such as granulations or warts, pitting, ridges, crests or folds and may also have spines or other processes and extensions (fig. 1, S). Presence or absence of a lorica, position and number of any spines and pattern of ornamentation are often diagnostic characters.

Foot

In many rotifers the body terminates in a post-anal section known as the foot (fig. 1, F); in some genera the foot is lacking and the anus is more or less terminal (e.g. figs 32-34). Presence or absence of a foot is a diagnostic feature which must be determined with particular care. In loricate rotifers the foot is usually retractile into the lorica through the *foot opening*; in non-loricate species it can usually be drawn up into the body. The foot may appear to consist of several sections or segments which can be telescoped into one another (fig. 20). In *Trichocerca* (fig. 10) it is very short and consists of only one or two such segments. In some genera it is long and very flexible, appearing rather wrinkled or ringed, as in *Brachionus* (fig. 1) and *Testudinella* (fig. 26). In most genera the foot terminates in one to four *toes*, usually two, which may be small to large, cone-, needle- or leaf-shaped, and equal or unequal in size. In *Testudinella* (fig. 26) the toes are replaced by a circle of cilia.

In some genera the foot is represented by a *stalk*. In the sedentary members of these genera, the stalk is easily recognized as that part of the rotifer which anchors it to the substrate; it has a holdfast in place of toes. A few stalked rotifers are not sedentary, however, but swim freely in the plankton. These are members of the genus *Collotheca*, which have slim stalks with or without holdfasts (fig. 27), or of the genera *Conochilus* and *Conochiloides* which have short stout stalks, by the lower ends of which they may be joined into a ball-like colony (figs 28, 120).

Locomotion and adhesion

Locomotion is usually effected either by the beating of the coronal cilia or by movements of the foot.

Most planktonic rotifers swim by using the cilia of the corona to propel themselves through the water, often rotating about the longitudinal axis as they do so. In colonies of *Conochilus*, the individuals are bound together by the gelatinous cases around their feet, and the whole spherical colony revolves through the water as a unit under ciliary action. Some swimming rotifers like *Brachionus* have a well-developed foot and appear to use it as a rudder, but swimmers without a foot, such as *Keratella*, apparently do not suffer any disadvantage.

Rotifers with toes use these to anchor themselves to a substrate. One or more pairs of *foot glands* (fig. 1, FG) discharge an adhesive secretion at the points of the toes. Species using the corona for both feeding and swimming anchor on a piece of plant or detritus to collect food particles from the water. Browsers on detritus particles also hold with the toes and creep forward using foot and corona alternately. In *Rotaria* and other members of the order Bdelloidea, a leech-like type of creeping gives the group its name. The two coronal discs present in this group can be tucked away under a *rostrum* or snout which then takes the lead in creeping (fig. 20). Telescoping of the foot and body segments with alternate extension and retraction allows the characteristic looping movement. For feeding or swimming, the rostrum is pushed back like a hood, and the coronal cilia are unfurled.

Extension and retraction of head, body or foot are made possible by a well-developed muscle system. Longitudinal muscles (fig. 1, M) permit retraction of foot or head up to and into the body. Extension is effected by arcs or circles of muscles around the body exerting pressure on the turgid body contents. Particularly well developed are the neck muscles, which form complete circles below the apical region and cause head extension, and can also close tightly over the head when it is retracted. Muscles hold together the separate plates of a lorica and allow expansion and contraction as head or foot are withdrawn or extended.

Many planktonic genera have no foot, although foot glands may still persist, as in *Synchaeta*, where their product is used to anchor the eggs to body or substrate. Movement in footless rotifers is either solely by the cilia of the corona as in *Keratella* (fig. 30), or by special appendages such as the 'blades' of *Polyarthra* (fig. 33) or the 'arms' of *Hexarthra* (fig. 34). Muscles extending to the tips of the arms in *Hexarthra*, or blocks of muscles acting against the cuticle in *Polyarthra*, allow rather jerky but rapid jumping movements.

Food, jaws and digestive system

The food may consist of small algae, detrital particles or bacteria collected by the coronal cilia or browsed from the surface of vegetation and larger detrital particles. Larger algae or small animals, including other rotifers, may be seized and engulfed whole, or pierced and sucked. Conspicuous stomach contents can be seen in many species, including *Ascomorpha* (fig. 36) with an extensive stomach filled with a green algal mass, and *Asplanchna* (fig. 35) in which plankters such as the rotifer *Keratella* and the diatom *Asterionella* can be easily recognized.

The mouth lies ventrally on the head and may open by way of a ciliated funnel-like depression. It leads to the pharynx containing the jaws or *trophi*. The pharynx region with jaws is often referred to as the *mastax*. The jaws consist of a number of hard cuticularized parts, basically seven, which retain their shape after death or compression,

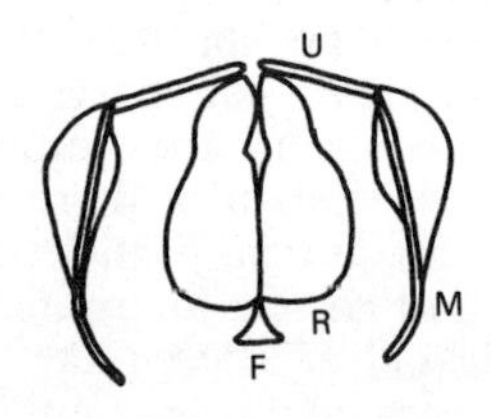

Fig. 2. Diagram of jaw parts. Ventral. F fulcrum; M manubrium; R ramus; U uncus.

though not necessarily their relative positions. The basic constituent parts are one *fulcrum* and a pair each of *unci*, *rami* and *manubria* (fig. 2). Several distinct types of jaws are usually recognized, each type characterized by the relative size and shape of the parts, and by their mode of functioning. The type of jaws forms the usual basis for division into families. In this key the jaws are used as diagnostic characters only in cases where other features do not present a clear choice, although notes on the jaws are included as confirmatory evidence.

Probably the easiest jaw type to recognize is the *malleate* type (fig. 3), which is characteristic of the Brachionidae. All parts are well developed and strong, the fulcrum is short, the rami broad. The unci have ridges across the surface terminating in teeth at the inner edge. The action is to cut and chew, between the teeth, particles collected by the cilia of the corona and mouth funnel. Another large jaw type is the *incudate* (fig. 4), characteristic of Asplanchnidae. Only the rami are well developed, being large and like forceps; other parts are small. The jaws lie reversed in the pharynx and are swivelled and pushed through the mouth, forceps-point forwards, to seize and hold prey.

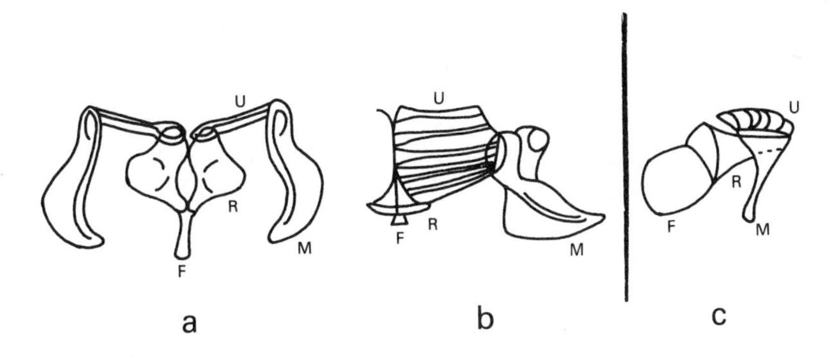

Fig. 3. Malleate jaw type. Jaws of *Brachionus calyciflorus; a,* ventral; *b,* dorsal; *c,* lateral.

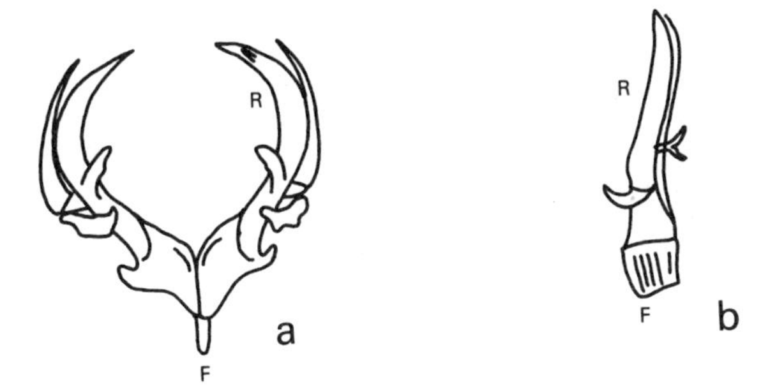

Fig. 4. Incudate jaw type. Jaws of *Asplanchna brightwelli; a,* ventral; *b,* lateral (after Beauchamp 1909).

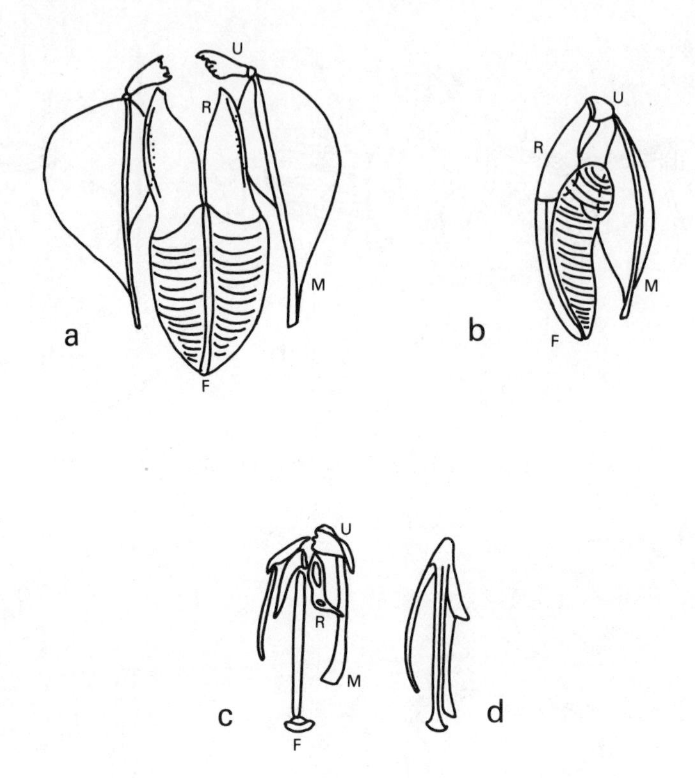

Fig. 5. Virgate jaw type. Jaws of: *a, b, Synchaeta oblonga; a,* ventral; *b,* lateral; *c, d, Trichocerca bicristata; c,* ventral; *d,* lateral.

The jaw types of other families tend to be more difficult to recognize, being often small and difficult to pick out from the mass of tissue in the head, especially in the living animal. The *virgate* type (fig. 5) is characteristic of Synchaetidae, Gastropodidae and Trichocercidae. All parts tend to be slender, with long fulcrum and manubria; the rami may be small. The fulcrum provides attachment for a hypopharynx muscle which gives the jaws a sucking action. This muscle is particularly well-developed in *Synchaeta* (fig. 5, *a,b,*), where the jaw mass is large and conspicuous. In the Gastropodidae and especially in the Trichocercidae, the jaws tend to be asymmetrical, in some species markedly so (fig. 5, *c,d,*). In these families the jaws are used for piercing; sucking then occurs by pumping of the hypopharynx muscle. The corona here is used mainly for swimming. In

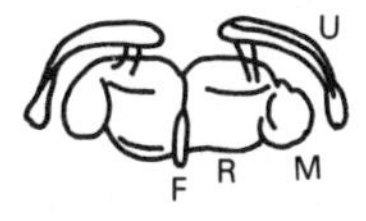

Fig. 6. Uncinate jaw type. Jaws of *Collotheca*.

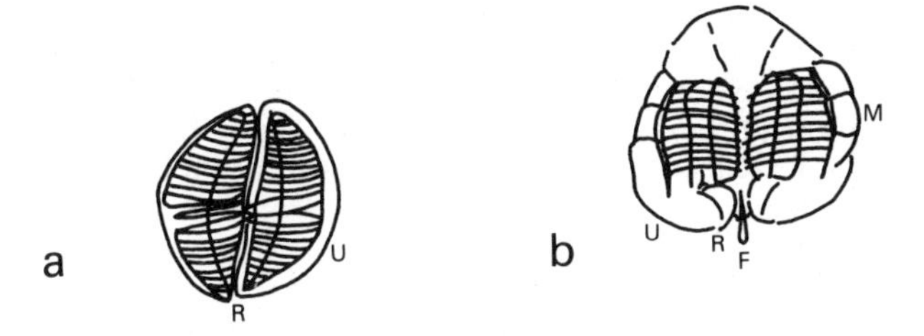

Fig. 7. Ramate (*a*) and malleoramate (*b*) jaw type. Jaws of *a, Rotaria neptunia,* ventral; *b, Filinia longiseta,* ventral.

the *uncinate* type (fig. 6) found in the Collothecidae, the jaws are small with broad rami. Prey captured by the long coronal cilia is taken into a voluminous pharynx. The *ramate* and *malleoramate* types (fig. 7) are found in *Rotaria* and other bdelloid rotifers and also in Conochilidae and Testudinellidae. The toothed unci are well developed; other parts are reduced, especially in the ramate type. Particles collected by coronal cilia are chewed by the unci.

Two or more salivary glands open into the pharynx. From the pharynx the food passes through the oesophagus to the stomach, which receives the openings of a pair of gastric glands. Most of the tissues in Rotifera are syncytia, lacking cell walls between the nuclei. However, except in the bdelloid rotifers, the stomach consists of discrete cells with granular inclusions. From the stomach, the intestine leads to the anus. The end of the intestine may be known as the cloaca, as it receives the ducts from the osmoregulatory and reproductive systems. In one or two genera, such as *Asplanchna* the intestine and anus are absent (fig. 35); unwanted hard remains are ejected through the mouth. Some genera, e.g. *Ascomorpha* (fig. 36), accumulate dark accretion bodies, probably of excretory material.

Osmoregulatory and excretory system

The osmoregulatory system (Pontin 1964, 1966) consists of a pair of *protonephridia* and a *contractile vesicle* or bladder (fig. 1). Each protonephridium is essentially a coiled tubule bearing at its distal end a number of *flame bulbs* (also called flame cells) (fig. 1), the number varying with the species. Within each flame bulb hangs a bunch of fused cilia forming the *flagellum*. During life this is in continuous motion, the waves of movement passing along the flagellum giving the impression of a flickering flame. Fluid from the body cavity passes through the bulb face into the bulb cavity and down a fine capillary tubule to join the main tubule. The proximal part of the main tubule is really a much-coiled channel in a syncytial mass of cytoplasm, which may be compact as in *Asplanchna priodonta* (fig. 35) or extensive as in *A. brightwelli* (fig. 89). Here ion exchange and excretion take place. The distal part of the protonephridial tubule enters the contractile vesicle, which is a round or pear-shaped bag with a muscle net and nerve ganglion. When full, the vesicle discharges to the outside through the cloaca. Regular filling and emptying of the vesicle ensures the removal of surplus fluid or urine.

Reproductive system and egg types

Rotifers of the order Monogononta, which includes most of the planktonic species, have only one ovary, but members of the order Bdelloidea have two. The small ovary is combined with a larger yolk gland in a common membrane which continues as the oviduct to the cloaca. The combined ovary – yolk gland may be referred to as the ovovitelline gland or germovitellarium. In *Asplanchna*, where the posterior intestine is absent, the oviduct and contractile vesicle have a common duct to the exterior (fig. 35). Eggs in the early stages of development can usually be seen in the yolk gland. The number of eggs which can be produced by the mother is determined by the fixed number of nuclei in the ovary, about 10-50.

Three types of eggs are produced in members of the Monogononta.

(1) *Asexual eggs*. Amictic females produce diploid parthenogenetic eggs which cannot be fertilized. They may be dropped, laid on plants or carried attached to the cuticle or foot; in a few cases they develop internally. Asexual eggs develop into either amictic or mictic females.

(2) *Sexual eggs*. Small haploid sexual eggs are produced by mictic females. If unfertilized, they develop into males, and a group of such eggs may be seen attached to the mother. If the sexual eggs are fertilized, restoring the diploid state, they become

(3) *resting eggs*. These become larger than the male eggs, acquiring a thick, often ornamented or spiny shell, and a store of food, often in the form of oil drops. They may be laid, or may develop inside the shell of the dead mother. They resist extremes of temperature and desiccation, remain dormant for periods ranging from several days to

months, and always develop into amictic females. Fertilization takes place by impregnation through the cuticle of the female, usually when she is small and young, and before the cuticle has hardened. Resting eggs may sometimes be seen developing inside a mother also carrying male eggs; in such a case, fertilization probably took place during the egg-bearing phase. Mictic and amictic females are usually indistinguishable in appearance, except for the eggs each may carry.

In planktonic and other monogonontic rotifers, sexual and asexual phases may alternate. In bdelloid and some monogonontic species, where males are unknown, reproduction is assumed to be entirely asexual. In the genus *Seison*, reproduction is exclusively sexual.

C. SEXUAL PERIODICITY

The asexual phase is characterized in many planktonic rotifers by the development of large numbers of individuals; by this means the species can take rapid advantage of favourable conditions. In some species maximum numbers appear in spring, in others in summer or autumn, while some have both spring and autumn outbursts. The exact timing and size of the maxima in each species vary between years and between habitats, depending on temperature, food availability and other environmental factors.

In many planktonic rotifers the asexual phase is followed, and may be completely terminated, by the development of sexual or mictic eggs and the appearance of males in the population. Males of *Brachionus*, for example, appear fairly frequently, and are seen more commonly than those of most other genera. The appearance of males is accompanied by that of fertilized resting eggs, which may over-winter and develop in spring. In some planktonic species, males may not always be seen. For example, *Keratella* species may pass through several or many seasons of asexual reproduction without males and resting eggs, especially in larger bodies of water. Some species may never reproduce sexually. Where no resting eggs are produced, the species survives winter in the form of slowly-reproducing amictic females. A perennial species is one which persists, largely in the form of parthenogenetic females, throughout the year and from one year to the next, in contrast with those species which are present in the adult form for certain seasons only.

The causes of the onset of the sexual phase by the development of mictic females have been the subject of research over many years, some of it with conflicting results. Temperature, diet and pH are among environmental factors investigated. A survey of the main work in this field and a discussion of numerous aspects of parthenogenesis and sexual reproduction in rotifers may be found in the excellent paper by Birky & Gilbert (1971).

D. POLYMORPHISM AND CYCLOMORPHOSIS

Collections of planktonic rotifers from different localities usually reveal a phenomenon common to many species, that of polymorphism. For example, one may collect some forms of *Keratella cochlearis* with long posterior spines, others from elsewhere with short spines or none, some with a smooth lorica, others with a rather warty or spiny lorica, some with a symmetrical pattern on the dorsal surface, others with an irregular pattern. Collections of the same species from one locality over a period of time will often show progressive changes in one or more characteristics from generation to generation. For example, *K. cochlearis* may be collected in spring as the long-spined form; successive collections from the same locality may show forms with shorter and shorter posterior spines until, by the summer, some may have no posterior spine at all. Changes do not occur during the lifetime of one individual, but only from one generation to the next. While increases or decreases in spine length are the most obvious changes, many species also show changes in body proportions or other features, such as humps or projections. In many localities, changes such as decrease in spine length from spring to summer may be repeated each year, with but slight variations in timing. A cyclic change of this type is known as cyclomorphosis.

Many of these different forms have appeared in the literature under the guise of genetic varieties or even of species, and have considerably added to the taxonomic confusion. However, as in the case of sexual reproduction in rotifers, factors in the environment have long been suspected of at least some responsibility for the appearance of these different forms. Many factors have been investigated, including temperature, food, pH and turbulence, again with some conflicting results (see Hutchinson (1967) for a summary). Ruttner-Kolisko (1974) summed up the effects of these factors for *Keratella cochlearis*, showing that it is the degree and the combination of these factors which determine the form occurring in any locality. For example, as the degree of eutrophication increases, so does the tendency to find short-spined forms; while as turbulence increases, so do the chances of finding the form with warted or spiny lorica.

It is probable that different environmental factors have different effects on different species, and may even differ in their influence upon different characteristics of the same rotifer. Progressive changes in the environment during the seasons, such as temperature, may be responsible for some of the cyclical changes in form.

E. MORPHOLOGY OF THE MALE (see fig. 8)

The male rotifer may be very similar in appearance to the female, but slightly smaller, with the various body structures well-developed, (e.g. *Rhinoglena*, fig. 125); or he may be extremely reduced in size and structure (e.g. *Polyarthra*, fig. 144). Lorica, foot, jaws, gut and contractile vesicle may be poorly represented or absent. However, sense organs, brain, muscle system, testis and some type of copulatory device are usually well-developed (see fig. 8).

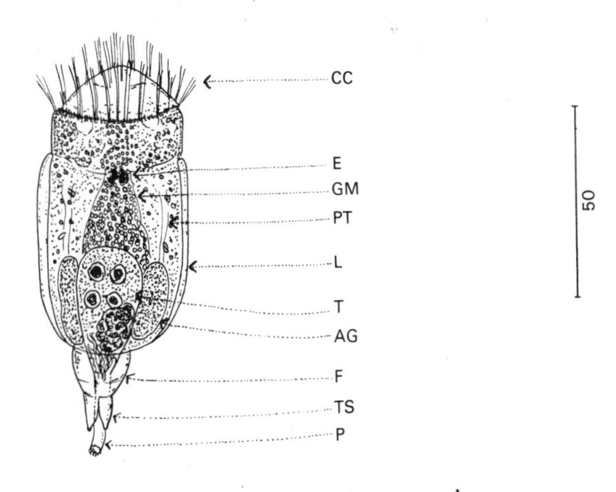

Fig. 8. *Brachionus angularis*. Male, ventral. AG accessory sexual glands; CC coronal cilia; E eye; F foot; GM mass of gut rudiment; L lorica; P penis; PT protonephridium; T testis; TS toes.

Head

The corona may be well-developed and differentiated as in the female (*Euchlanis*, fig. 136) or be only a ciliary circle. Sensory antennae are usually present and well-developed, the brain is large and there is usually a large red eye or eyes. A few genera have an apical snout or process, which may be present in the female as in *Rhinoglena* (fig. 125) or absent as in *Filinia* (fig. 126).

Body

The males of loricate females usually have a lorica, which is, however, poorly developed, thin and delicate, and may show merely as lines or furrows in the cuticle. Spines and projections of the lorica, found in the female, are rarely, if ever, present in the male (figs 128, 131).

Foot

Males of females with a foot usually have a foot also, frequently short. The males of some *Trichocerca* (fig. 140) and of *Gastropus* species (fig. 141) are footless. Toes are usually present on the foot, although in some species the foot ends in bristles or in a circle of cilia (*Testudinella* fig. 129, as in female). In some males (*Gastropus, Collotheca, Conochilus*; figs 141, 145, 127) there is no differentiation between the foot and the terminal part of the body.

Digestive system

Rhinoglena (fig. 125) is the only genus in which the male possesses a fully-formed and probably functional digestive system. The jaws resemble those of the female. In *Asplanchnopus* the male digestive system is well-developed, but probably functionless (fig. 134). In all other males the gut and jaws are strongly reduced; they may be represented by a strand of tissue often supporting the testis (*Asplanchna priodonta,* fig. 139*a*), or by a granular or globular mass (*Asplanchna brightwelli,* fig. 139*b*), or they may be totally absent (*Keratella,* fig. 131).

Reproductive system

The single testis is usually large, round or pear-shaped, supported by the rudimentary gut strand where present, and often surrounded by oil drops, and containing sperm. It opens by a sperm duct, often ciliated. Accessory sexual glands may be present. Copulation is effected in one of three ways: (a) by an external penis, which is usually large and conspicuous (e.g. *Brachionus,* figs 8, 128) though sometimes short (*Synchaeta*, fig. 135), lying dorsal to the foot which it often dwarfs; the sperm duct, which passes through the penis to open at its tip may be stiffened at its distal end as in *Brachionus* (fig. 128); the penis can be retracted and protruded; (b) while there is no actual penis, the sperm duct can be everted (turned inside out) for copulation (e.g. *Asplanchna,* fig. 139); (c) the posterior portion of the body is drawn out or protruded during copulation to form a 'penis' (e.g. *Filinia,* fig. 126).

In most species, fertilization is probably effected by impregnation through the soft cuticle of a young female.

F. CLASSIFICATION

The class Rotifera or Rotatoria is usually included in the phylum Aschelminthes or Nemathelminthes, together with the Nematoda, Gastrotricha and other groups. The class is divided into three orders, Monogononta, Bdelloidea, and Seisonidea.

The order Monogononta includes all the true planktonic rotifers, plus many semi- and non-planktonic species, and some sedentary ones. Many members of the order occur in fresh waters, some in brackish waters, and some in the sea. They are characterized by the possession of one ovary and by reproduction which is either entirely asexual or has asexual and sexual phases, with males which are more or less reduced.

Members of the order Bdelloidea usually show leech-like or looping-creeping movement. They are almost always found in mud, vegetation or detritus and only very occasionally do any appear in the plankton. There are two ovaries in each individual, and reproduction is entirely asexual, no males having been found.

The order Seisonidea contains only one genus, exclusively marine and epizoic. Males and females are similar in appearance and size; gonads are paired and reproduction is exclusively sexual.

The classification given in Table 1 is based on that of Voigt (1957).

TABLE 1

CLASSIFICATION OF ROTIFERA

	Genera with some planktonic or semi-planktonic members	Littoral or periphytic genera, sometimes collected in plankton	Littoral, periphytic or benthic genera rarely or never in plankton
Order **MONOGONONTA** Suborder PLOIMA Family BRACHIONIDAE	*Rhinoglena, Epiphanes, Brachionus, Euchlanis, Anuraeopsis, Keratella, Argonotholca, Kellicottia, Notholca, Squatinella.*	*Platyias, Mytilina, Lepadella, Colurella.*	*Proalides, Microcodides, Cyrtonia, Macrochaetus, Trichotria, Eudactylota, Wolga, Lophocaris, Diplois, Tripleuchlanis, Vanoyella, Dipleuchlanis, Pseudonotholca, Paracolurella.*

TABLE I (continued)

CLASSIFICATION OF ROTIFERA

	Genera with some planktonic or semi-planktonic members	Littoral or periphytic genera, sometimes collected in plankton	Littoral, periphytic or benthic genera rarely or never in plankton
Family LECANIDAE		*Lecane.*	*Bryceella, Tetrasiphon, Proalinopsis, Proales.*
Family LINDIIDAE			*Lindia.*
Family BIRGEIDAE			*Birgea.*
Family NOTOMMATIDAE		*Cephalodella.*	*Drilophaga, Scaridium, Monommata, Tylotrocha, Dorystoma, Itura, Enteroplea, Sphyrias, Pseudoharringia, Metadiaschiza, Taphrocampa, Eothinia, Rousseletia, Notommata, Wulfertia, Resticula, Eosphora, Pleurotrocha.*
Family TRICHOCERCIDAE	*Trichocerca.*		*Hertwigella, Elosa.*
Family GASTROPODIDAE	*Gastropus, Ascomorpha.*		
Family DICRANOPHORIDAE			*Balatro, Albertia, Pedipartia, Aspelta, Streptognatha, Erignatha, Wierzejskiella, Myersinella, Wigrella, Encentrum, Paradicranophorus, Dicranophorus, Dorria.*
Family ASPLANCHNIDAE	*Asplanchnopus, Asplanchna.*		*Harringia.*
Family SYNCHAETIDAE	*Polyarthra, Synchaeta, Ploesoma.*		
Family MICROCODONIDAE			*Microcodon.*

Suborder FLOSCULARIACEA			
Family TESTUDINELLIDAE	*Testudinella, Pompholyx, Hexarthra, Filinia, Tetramastix, Fadeewella, Trochosphaera, Horaella.*		
Family FLOSCULARIIDAE (all sedentary)			*Limnias, Floscularia, Octotrocha, Sinantherina, Lacinularia, Beauchampia, Ptygura.*
Family CONOCHILIDAE	*Conochiloides, Conochilus.*		
Suborder COLLOTHECACEA			
Family COLLOTHECIDAE (nearly all sedentary)	*Collotheca.*		*Stephanoceros, Cupelopagis, Atrochus, Acyclus.*
Order **BDELLOIDEA**			
Family HABROTROCHIDAE			*Otostephanus, Scepanotrocha, Habrotrocha.*
Family ADINETIDAE			*Adineta, Bradyscela.*
Family PHILODINIDAE		*Rotaria.*	*Zelinkiella, Mniobia, Anomopus, Didymodactylus, Ceratotrocha, Macrotrachela, Embata, Philodina, Dissotrocha, Pleuretra.*
Family PHILODINAVIDAE			*Abrochtha, Philodinavus, Henoceros.*
Order **SEISONIDEA**			
Family SEISONIDAE			*Seison.*

G. SYNONYMY

The nomenclature of the Rotatoria has been bedevilled by a multiplicity of synonyms of both generic and specific names. Names of many species have changed several times since they were first described. The most important of the generic synonyms are given in Table 2, while those for names of species are included in the keys. The original descriptions of each genus and type species are also referred to in the keys, and listed in the references under their respective authors.

TABLE 2

MAIN SYNONYMS OF GENERA

Old name	Name in current usage
Anapus	*Ascomorpha*
Anuraea	*Keratella*
Bipalpus	*Ploesoma*
Chromogaster	*Ascomorpha*
Colurus	*Colurella*
Dapidia	*Euchlanis*
Dinops	*Harringia*
Diplacidium, Diplax	*Mytilina*
Diurella	*Trichocerca*
Floscularia (in part)	*Collotheca*
Hydatina	*Epiphanes*
Mastigocerca	*Trichocerca*
Metopidia	*Lepadella*
Monostyla	*Lecane*
Noteus	*Platyias*
Notholca (*longispina* only)	*Kellicottia* (*longispina*)
Notops	*Epiphanes*
Pedalia, Pedalion	*Hexarthra*
Postclausa	*Gastropus*
Pterodina	*Testudinella*
Rattulus	*Trichocerca*
Rhinops	*Rhinoglena*
Sacculus	*Ascomorpha*
Salpina	*Mytilina*
Schizocerca	*Brachionus*
Stephanops	*Squatinella*
Triarthra	*Filinia*

H. METHODS OF COLLECTION AND STUDY

Planktonic rotifers are best collected by using a fine-mesh nylon or silk net on a pole or rope. A suitable mesh size is 180 meshes/inch (70/cm), though even this fineness may allow the passage of very small rotifers, including some males. Open water of lakes and ponds will yield the truly planktonic species. Collection, whether by accident or design, in the littoral region and around aquatic vegetation will result in a number of semi-planktonic and littoral species being encountered. Many species will live after collection for several days or longer if kept in a fairly cool place, if necessary in the least cold part of a refrigerator.

The ideal, and often the easiest, method is to examine the animals alive as soon as possible after collection. If the jar containing the animals is illuminated by a lamp from one side, many planktonic rotifers will quickly gather at that side. A small drop may then be taken from this region by pipette, and examined on a slide under the microscope. Observation of the living unconstricted animal is usually necessary to determine body shape, locomotion, presence or absence of a foot and other characters. However, many species may swim and turn too rapidly for comfortable observation. Preservation without narcotization often causes retraction of head and foot, and species with thin cuticles may become unrecognizable blobs. The animals can usually be studied by restricting the water on the slide to a small drop and gently adding a coverslip. Cotton wool strands or methyl cellulose can be added to slow down rapid swimmers. Where enough rotifers are present, the placing of the coverslip over a drop of collection usually results in specimens being trapped in various attitudes, some presenting ventral views, others dorsal, some retracted, others extended, so that the desired characters can be seen. However, it is as well to check that only one species is present, as similar species frequently occur together.

Examination of preserved animals can be a useful and sometimes essential adjunct to a study of living ones. On the preserved animal, the shape of a lorica, number of spines or shape of foot opening can be checked. Preservation in an extended condition can sometimes be achieved by using very dilute formalin (1%) or alcohol; 5% formalin is a suitable long-term preservative, especially for the loricate species. Many narcotics are not readily taken up by rotifers, which have a fairly impervious cuticle. Two effective narcotics are physostigmine salicylate (5×10^{-5} g mol^{-1}) and tricaine methane sulphonate (MS – 222). Stains are not very easily taken up, but a mixture of polyvinyl alcohol, lactophenol and lignin pink makes a suitable stain and mount to show the jaws. Satisfactory permanent preparations must be in liquid with ringed mounts. Instructions for narcotization and mounting are given by Hanley (1949, 1954).

The difficulty of making and maintaining permanent preparations has prevented the formation of an adequate type collection of rotifers. Drawings and notes are therefore of particular importance for reference. Possibly a collection of photomicrographs might form the basis of a modern 'type collection'.

I. REFERENCE WORKS

A general account of the anatomy and physiology of the Rotifera is given in *The Invertebrates* Vol. 3 by L. H. Hyman (1951), and the survey by Josef Donner has been published in English as *Rotifers* (Donner, 1966). Some elegant short papers by E. D. Hollowday (1945-50), describing different genera, were published in *Microscope*. *The Rotifera; or wheel-animalcules*, by C. T. Hudson and P. H. Gosse (1886, 1889), is beautifully written and illustrated, though taxonomically out-of-date.

The most important key to European rotifer species is *Rotatoria: die Rädertiere Mitteleuropas* by Max Voigt (1957), part of which is currently being revised. A key to the European planktonic species with information on biology and ecology is given by Agnes Ruttner-Kolisko (1974).

References to papers dealing with particular genera and species are given in the appropriate sections.

J. DISTRIBUTION AND RECORDS

This key includes all known British and Irish planktonic species. Records of these species were compiled from a number of sources, including Galliford (1950, 1953, 1954a, b, 1960, 1967, 1971, 1974 and personal communication), Horkan (personal communication), Hudson & Gosse (1886, 1889), *Limnofauna Europaea*, edited by Illies (1967); M. Doohan, A. Duncan, J. Green and E. D. Hollowday (all personal communications); and my own data. Initially it was intended to include records, by counties, in the notes for each species. Such records, however, tend to reflect the distribution of collectors, rather than of animals, and it was felt that they might give a misleading impression of the distribution of what are, probably, mostly widespread species. Mention of recorded occurrence is therefore only made where a species seems to be rare. Records so far collected now form the nucleus of a British list, and it would be valuable to have additions to this list.

KEY TO GENERA (FEMALES)

(See Scope of the Key, p. 5)

1 Foot present (figs 9-28); foot may be short or long, one-piece or of two or more telescopic segments, smooth or wrinkled, with or without toes— **2**

Note: foot may be retracted up into body; observe in both active and stationary animals. In loricate rotifers (see below) with retracted foot, the foot opening is usually observable. *Pompholyx* has a 'cloacal opening' but no foot (fig. 110), *Anuraeopsis* carries eggs attached to an anal appendage or egg-carrier (fig. 51*b*↙), which may appear foot-like when extruded from the lorica.

— Foot absent (figs 29-38)— **22**

2 Foot with toes (figs 9-25)— **3**

— Toes absent, foot ends in holdfast (figs 27, 121*d*), or in a circle of cilia (figs 26, 108*c*) or neither (fig. 28)— **20**

3 Lorica present, i.e. cuticle thickened to form a stiffish shell or lorica, somewhat flexible but retaining shape well during movement and after preservation, often bearing spines and warts, striations, ridges, etc. (e.g. figs 1, 9, 10, 11)— **4**

Note: difficulties may arise here in placing a loricate rotifer with a thin lorica e.g. *Gastropus* (fig. 16); note that a lorica usually persists in characteristic shape after the body contents have dispersed, and can be seen to be a separate part of the animal.

— Lorica absent, i.e. cuticle thin, very transparent, very flexible, allowing considerable variation in shape during movement, not usually retaining shape well after death or preservation (e.g. figs 23, 24, 25)— **15**

4 Lorica flattened dorsoventrally, or not flattened (e.g. figs 1, 9, 11, 69)— **5**

— Lorica more or less flattened laterally (e.g. figs 15, 70, 80)— **11**

5 Foot long, very flexible, worm-like, appearing wrinkled, fully retractile (fig. 1). Lorica usually more or less flattened dorsoventrally (*Brachionus calyciflorus* sack-shaped, fig. 1), usually stout; anterior end almost always with 2, 4 or 6 spines; posterior end rounded, angled or with 1 or 2 spines; foot opening posterior, with or without spines. (Figs 42-47)— BRACHIONUS (p. 50)

— Foot fairly short, of 1-4 sections, not wrinkled (figs 9, 10)— **6**

6 Lorica more or less cylindrical (figs 9, 10)— **7**

— Lorica more or less flattened dorsoventrally (figs 11, 48, 69)— **8**

7 Head with conspicuous transparent, semi-circular, non-retractile head-shield (figs 9, 65*a*↙). Lorica cylindrical with 3 posterior spines; body and jaws symmetrical (fig. 65)— SQUATINELLA (p. 80)

— No head-shield (but see *Trichocerca capucina* with hood-like projection from lorica, fig. 71). Lorica straight or humped, with or without crests and anterior spines; body usually showing asymmetry especially in jaws and position of spines and eye; 1 or 2 toes; if 2 often unequal in size, needle-shaped, often twisted over each other; toes often accompanied by 1 or more spines or bristles (substylus) (figs 10, 71-79)— TRICHOCERCA (p. 86)

8(6) Lorica with spines on anterior border and at posterior corners (lorica resembles that of *Brachionus* fig. 44), covered with tiny spinelets. Foot of 3 sections, only partly retractile, toes short (figs 11, 67)— PLATYIAS (p. 82)

— Lorica oval to shield-shaped, often without spines, or with short spines at anterior corners. Foot of 1 or more sections (figs 13, 14, 48)— **9**

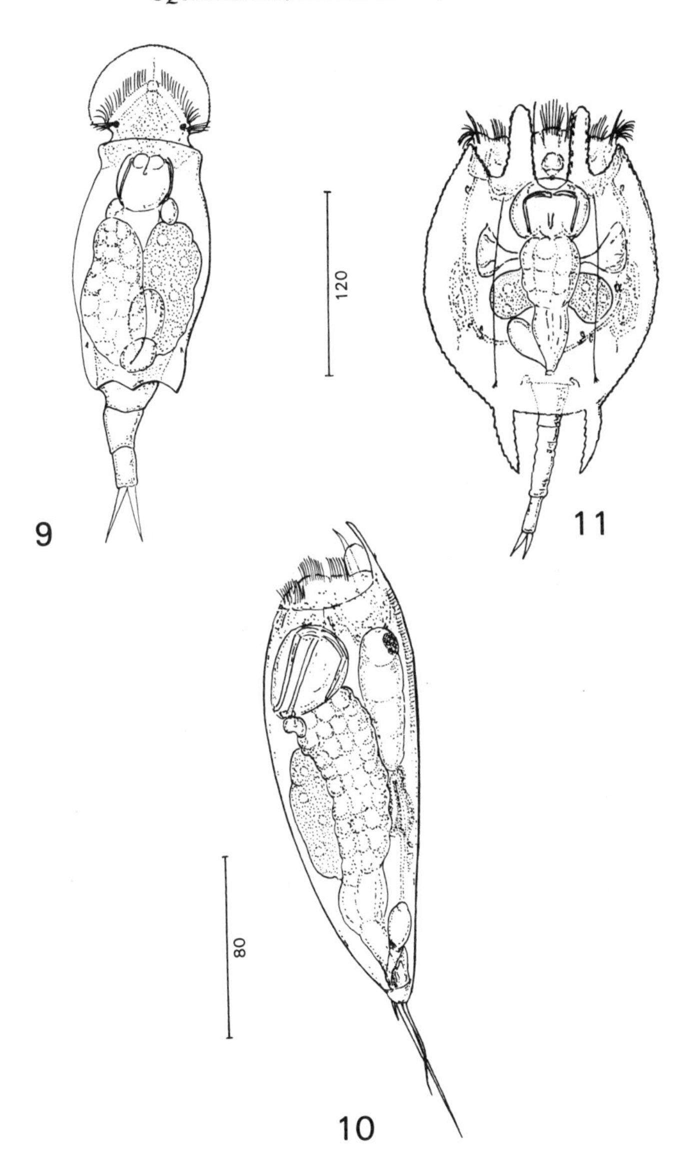

Figs 9-11. 9, *Squatinella tridentata*. Female, dorsal. 10, *Trichocerca similis*. Female, lateral. 11, *Platyias quadricornis*. Female, dorsal.

9 Lorica with more or less flattened ventral surface, dorsal surface convex (figs 12, 48), with or without dorsal crest and lateral extensions or wings; outline oval, egg- or pear-shaped; dorsal plate with deep indentation posteriorly. Foot short, 2-3 sections; 2 toes, large, sword- or spindle-shaped. One eye— EUCHLANIS (p. 61)

— Lorica more or less flattened dorsally and ventrally, of 2 plates or undivided (figs 69, 106)— **10**

10 Lorica oval, pear- or shield-shaped, with dorsal and ventral plates (figs 13, 106); anterior opening of lorica broad and shallow, with lateral edges prolonged into angles or short spines; posterior end rounded or extended into a process. Foot very short, 1- or 2-piece; toes long, 2 or fused to 1. One eye— LECANE (p. 124)

— Lorica oval, egg- or pear-shaped, of 1 piece, with or without dorsal crests and lateral wings (figs 14, 69); anterior opening narrow, semi-circular. Foot opening deeply incised, foot 3-4 sections but only distal part and toes extend beyond lorica; 2 toes, short to long, pointed. Two lateral eyes— LEPADELLA (p. 84)

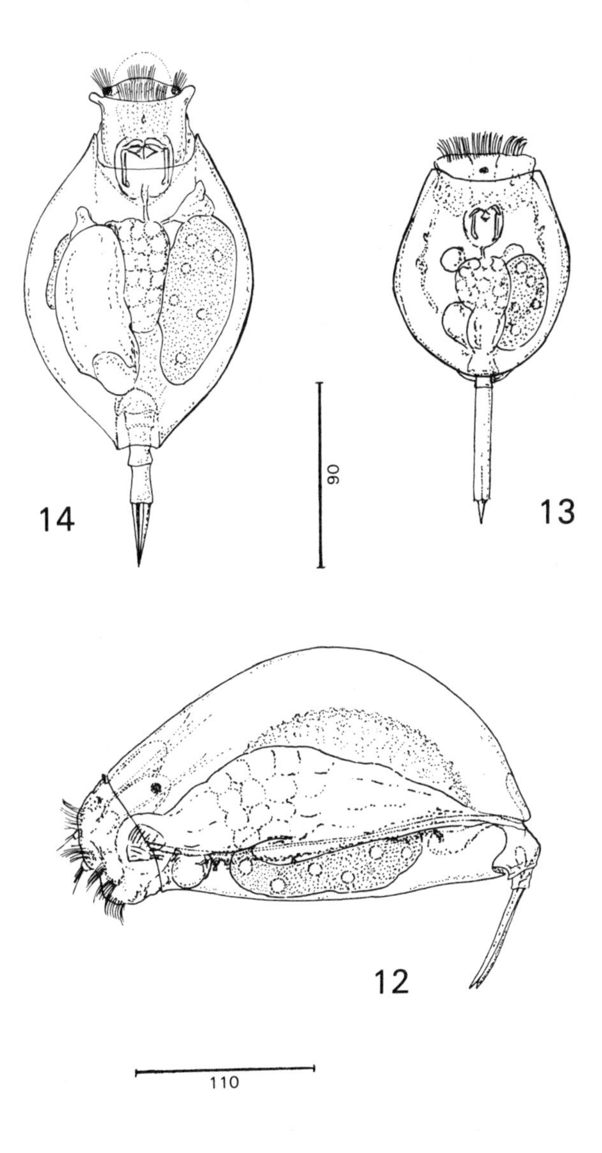

Figs 12-14. 12, *Euchlanis dilatata*. Female, lateral. 13, *Lecane lunaris*. Female, dorsal. 14, *Lepadella ovalis*. Female, dorsal.

11(4) Foot wrinkled; foot opening ventral; 1 or 2 toes (figs 15, 16)— **12**

— Foot not wrinkled, of 1 or more sections; foot opening posterior; 2 toes (figs 17, 18, 19)— **13**

12 Lorica stout, sack- or cone-shaped, ornamented with indentations, grooves or ridges (fig. 15)— PLOESOMA (p. 122)

— Lorica thin and delicate, flask-shaped, smooth (fig. 16)— GASTROPUS (p. 95)

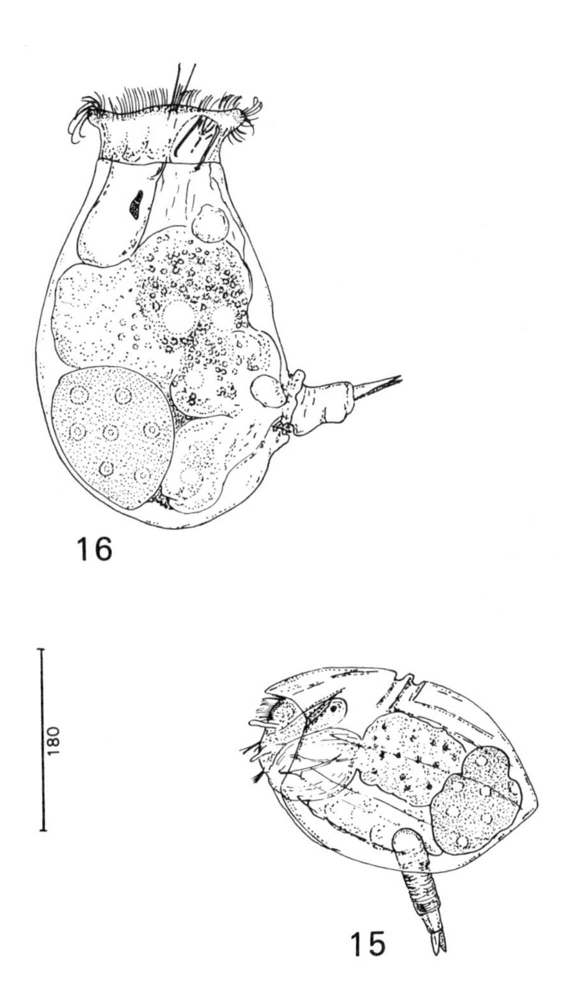

Figs 15-16. 15, *Ploesoma truncatum*. Female, lateral. 16, *Gastropus hyptopus*. Female, lateral. (15, after Hyman 1951 and others).

13(11) Body slightly curved towards ventral side (fig. 17); lorica delicate, of several pieces, with lateral cleft. Foot short; toes short to long—
CEPHALODELLA (p. 125)

— Lorica rather mussel- or purse-shaped in side-view (figs 18, 19)— **14**

14 Lorica fairly stout, with dorsal cleft (figs 18, 68), posterior and sometimes anterior corners of lorica with spines. Foot of 1 or more sections; toes fairly long, sword-shaped or slightly curved—
MYTILINA (p. 83)

— Lorica rounded or truncated anteriorly, posteriorly rounded or extended into a point; head with small retractile head-shield, appearing hook-like in side view (fig. 19). Foot of 3-4 sections—
COLURELLA (p. 85)

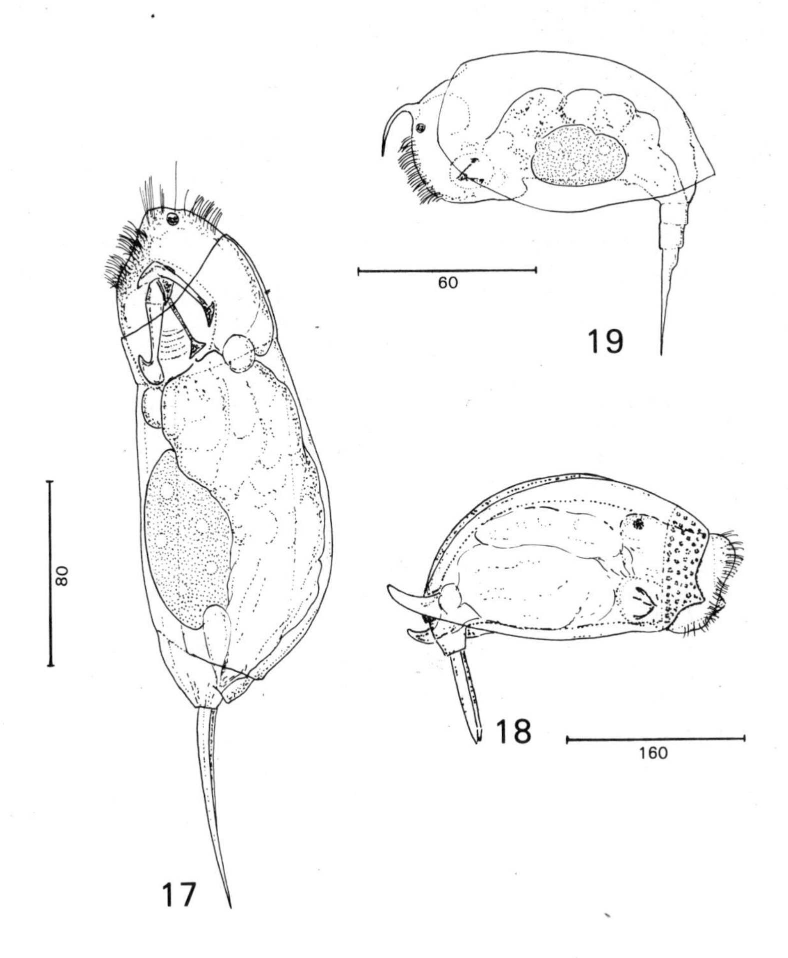

Figs 17-19. 17, *Cephalodella gibba*. Female, lateral. 18, *Mytilina ventralis*. Female, lateral. 19, *Colurella adriatica*. Female, lateral.

15(3) Capable of locomotion with corona retracted, by leech-like creeping with telescoping of body and foot sections (in addition to ciliary swimming with extended corona); corona of 2 adjacent circles of cilia (fig. 20). Extended foot long or very long, with 3 toes and 2 spurs (fig. 20)— ROTARIA (p. 146)

— Not capable of leech-like creeping, locomotion by coronal cilia (may be use of foot, but not in leech-like manner); corona not of 2 adjacent circles of cilia. Foot short to medium length, 2 toes, no spurs (figs 21-25)— **16**

16 Head with dorsal snout-like process (figs 21, 39*a*↙), corona with conspicuous large cirri or bristles extending in a V-shape from ventral mouth along sides of head, with smaller cilia extending on to snout— RHINOGLENA (p. 46)

— Head and corona not like this (figs 22-25)— **17**

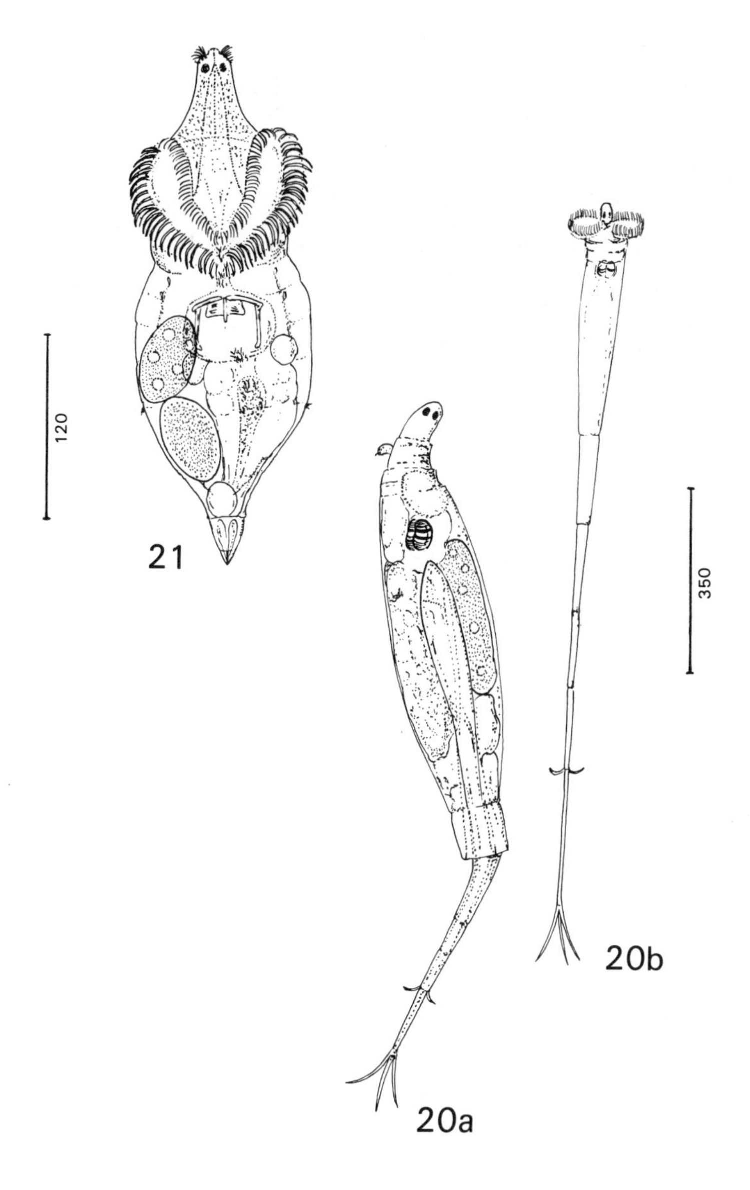

Figs 20-21. 20, *Rotaria neptunia*. Female, *a*, corona retracted; *b*, extended. 21, *Rhinoglena frontalis*. Female, ventral.

17 Posterior intestine and anus present (figs 23-25). Foot terminal— **18**

— Posterior intestine and anus absent (fig. 22). Large sack-like body, small ventral foot— ASPLANCHNOPUS (p. 100)

18 Head with 2 lateral ear-like auricles (fig. 23) and usually with 4 stiff sensory bristles— SYNCHAETA (p. 116)

— Head without auricles (figs 24, 25)— **19**

19 Jaws malleate, for chewing (fig. 40)— EPIPHANES (p. 48)

— Jaws incudate, for piercing (fig. 91*c*)— HARRINGIA (p. 109)

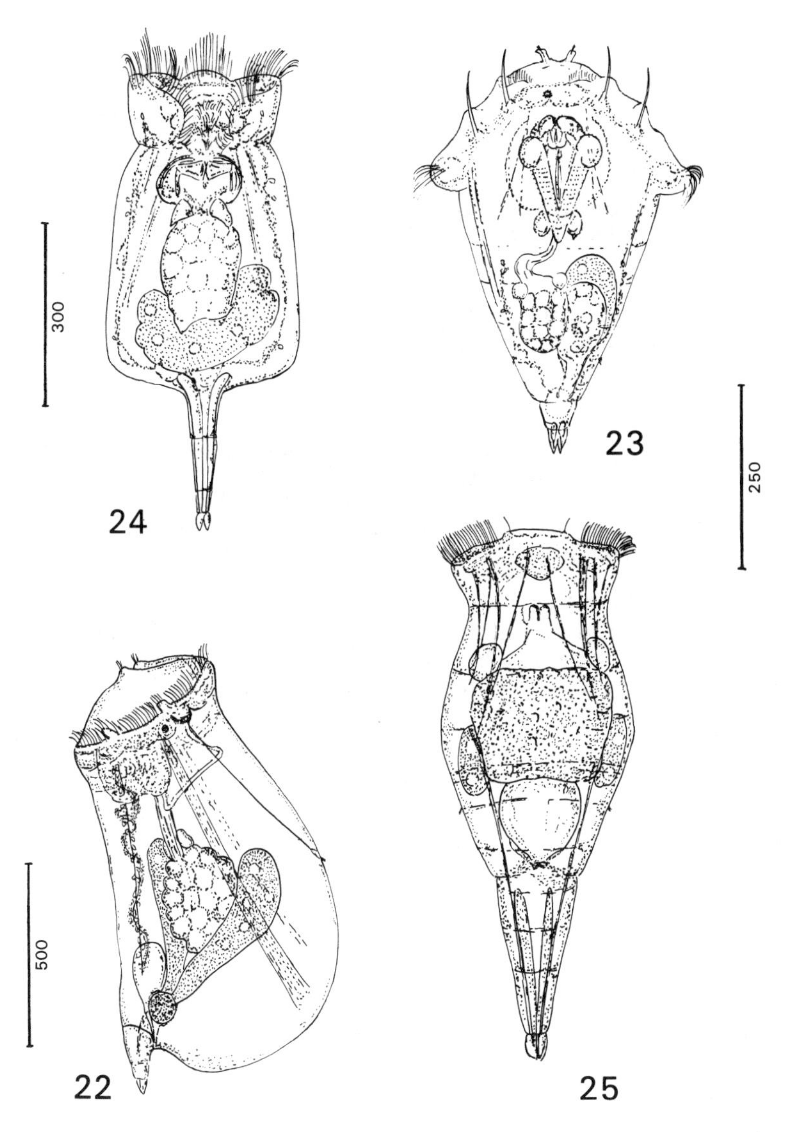

Figs 22-25. 22, *Asplanchnopus multiceps*. Female, lateral. 23, *Synchaeta pectinata*. Female, dorsal. 24, *Epiphanes brachionus*. Female, ventral. 25, *Harringia eupoda*. Female, dorsal. (22, after various authors, 25, after Beauchamp 1912).

20(2) Cuticle thickened to form a stiff shell or lorica, strongly flattened dorsoventrally, circular or oval in outline, foot opening ventral (figs 26, 108)— TESTUDINELLA (p. 126)

— No lorica, cuticle thin and flexible. Usually a gelatinous transparent case around foot and part of body (fig. 27). May be single animals or in ball-like colonies (fig. 120*a*)— **21**

21 Head more or less funnel-like with mouth (fig. 121*a*↙) in centre; corona circular or with 1 or 2 lobes, coronal cilia in single row, cilia of lobes often long. Foot long, slender, stalk-like, with or without holdfast (fig. 121*d*). Single animals in very transparent, often wide, gelatinous case (figs 27, 121)— COLLOTHECA (p. 142)

— Head not funnel-like, roughly convex (fig. 120); corona horseshoe-shaped, interrupted ventrally, not lobed, coronal cilia in double row; mouth (fig. 120*b*, *c*↙) in middle of corona or near dorsal edge of corona. Foot fairly stout, without holdfast (fig. 120). Solitary or in spherical colony (individuals separate in preserved material) (figs 28, 120)— CONOCHILOIDES & CONOCHILUS (p. 138)

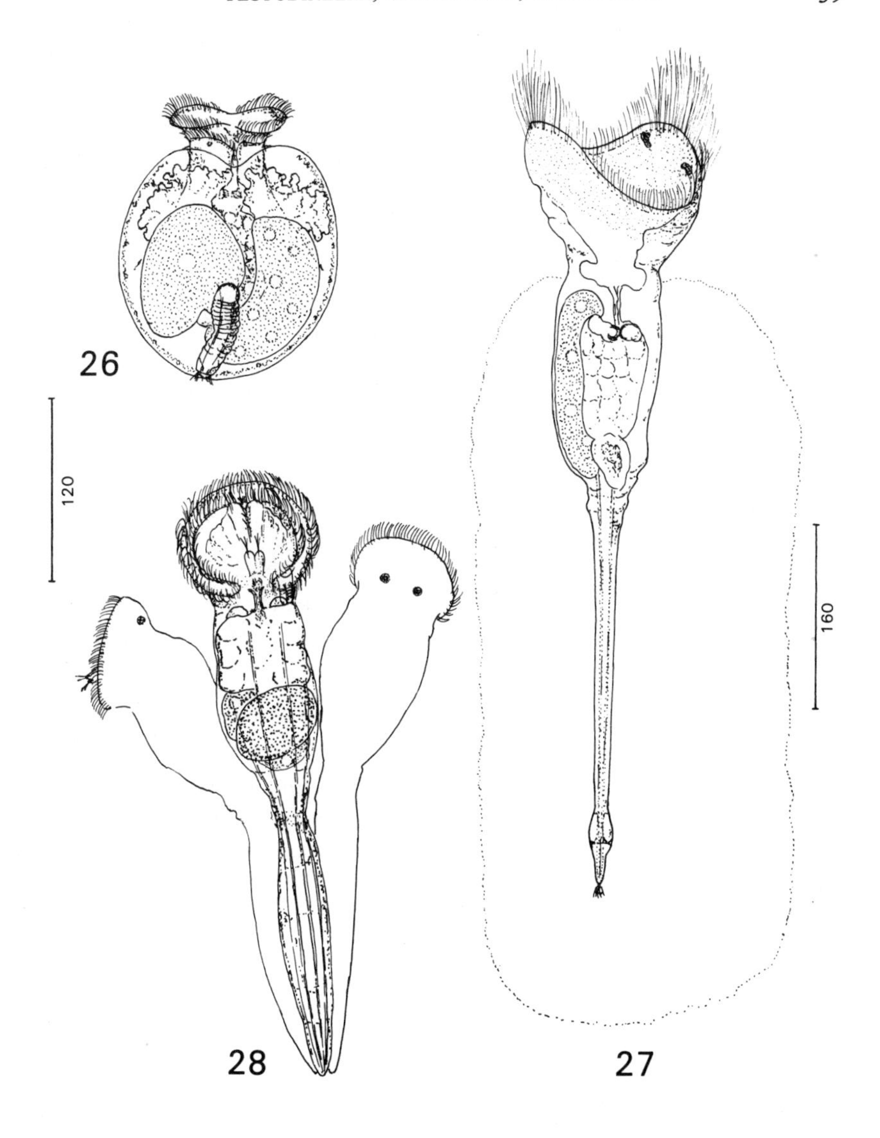

Figs 26-28. 26, *Testudinella patina*. Female, ventral. 27, *Collotheca mutabilis*. Female, lateral. 28, *Conochilus unicornis*. Female, 3 individuals of colony, centre one ventral. (28, partly after Hudson & Gosse 1886).

22(1) Body with spines or with blade-like or arm-like processes (figs 29-34)— **23**

— Body without these (figs 35-38)— **28**

23 Cuticle thickened to form a stiffish shell or lorica, somewhat flexible but retaining shape well during movement and after preservation, often ornamented with ridges or striations (figs 29-31)— **24**

— Cuticle thin, very transparent, very flexible, allowing considerable variation in shape during movement, not usually retaining shape well after death or preservation (figs 32-34)— **26**

24 Lorica with 6 anterior spines of almost equal length, with or without posterior spines (figs 30, 31)— **25**

— Lorica with 4 or 6 anterior spines of very unequal length, 1 posterior median spine (fig. 29)— KELLICOTTIA (p. 79)

25 Lorica marked on dorsal surface with a net-like pattern of lines (fig. 52)— KERATELLA (p. 66)

— Lorica marked on dorsal surface with longitudinal lines (fig. 31)— NOTHOLCA & ARGONOTHOLCA (p. 75)

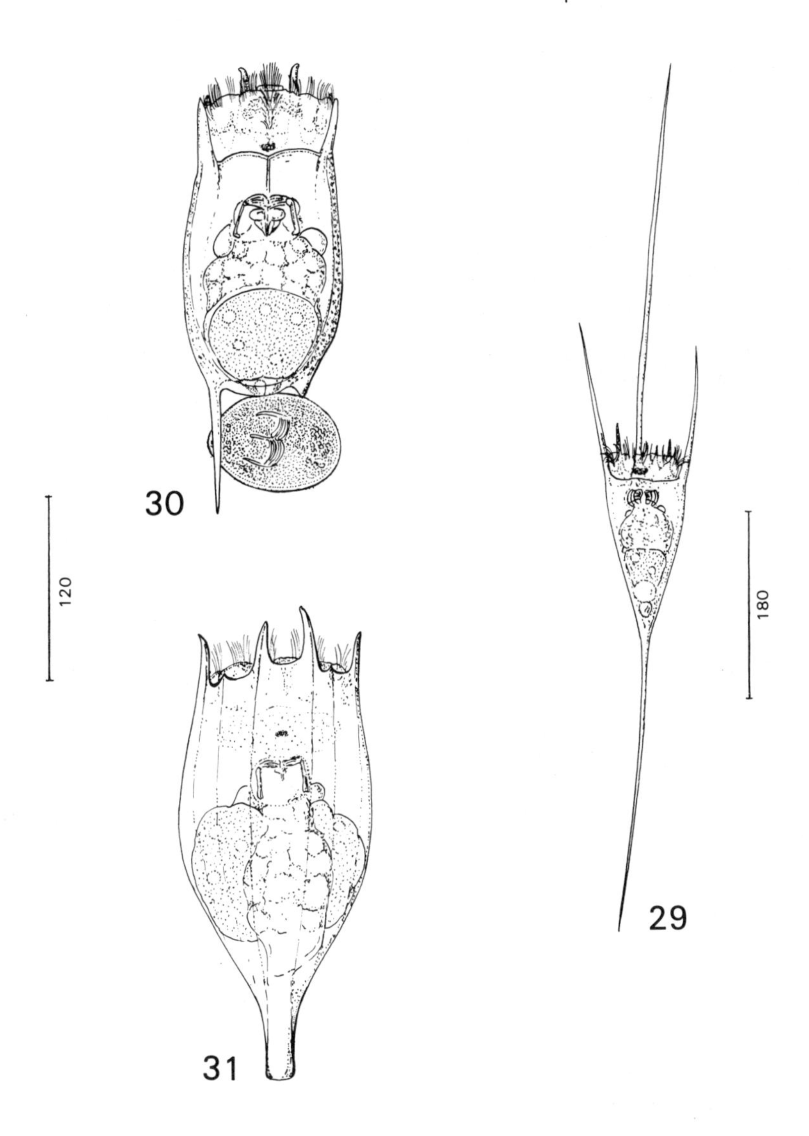

Figs 29-31. 29, *Kellicottia longispina*. Female, ventral. 30, *Keratella valga*. Female, ventral. 31, *Notholca labis*. Female, dorsal.

26(23) Body with 3 long or short movable spines or bristles, 2 anterior and 1 posterior (fig. 32)— FILINIA (p. 132)

— Body with 12 feather- or sword-like 'blades' (fig. 33) or with bristle-bearing 'arms' (fig. 34)— **27**

27 With 12 blades in 4 groups of 3, 2 groups dorso-lateral and 2 ventro-lateral (fig. 33)— POLYARTHRA (p. 109)

— With bristle-bearing arms (fig. 34)— HEXARTHRA (p. 130)

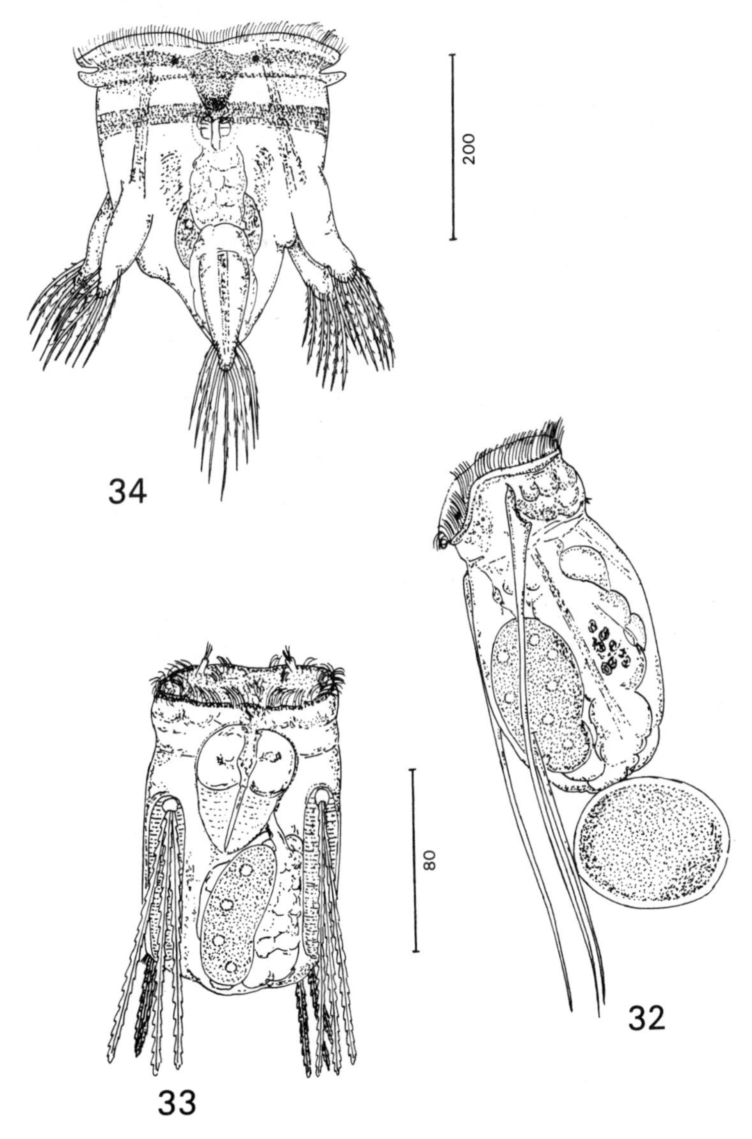

Figs 32-34. 32, *Filinia passa*. Female, lateral. 33, *Polyarthra remata*. Female, ventral. 34, *Hexarthra mira*. Female, dorsal. (32, after Hollowday personal communication).

28(22) Intestine and anus absent, stomach without extensions or accretion bodies (fig. 35). Large, very transparent, viviparous— ASPLANCHNA (p. 102)

— Intestine and anus present, though may be obscured by large stomach— **29**

29 Intestine obscured by large lobed stomach filling most of body cavity, often with conspicuous greenish gut contents and with 1 or more dark accretion bodies (figs 36, 83*a*↙)— ASCOMORPHA (p. 96)

— Stomach not like this; no accretion bodies— **30**

30 Lorica (i.e. shell of thickened cuticle) egg-cup- or shield-shaped (fig. 51) with fold down sides where dorsal and ventral plates meet (fig. 37). Often with large tear-drop-shaped eggs, attached to an egg-carrier (fig. 51*b*↙) and trailed behind during swimming— ANURAEOPSIS (p. 64)

— Lorica oval or shield-shaped; may be flattened dorsoventrally (fig. 109*b*) or 4-lobed in cross-section (fig. 110*b*). Often with round eggs attached to a secreted retractile thread (fig. 110*e*). ('Cloacal' opening not a foot opening)— POMPHOLYX (p. 128)

Note: Blade-less forms of *Polyarthra* representing first generation from the sexual egg will fail to key out satisfactorily here — see *Polyarthra* (p. 110) and fig. 95*d*.

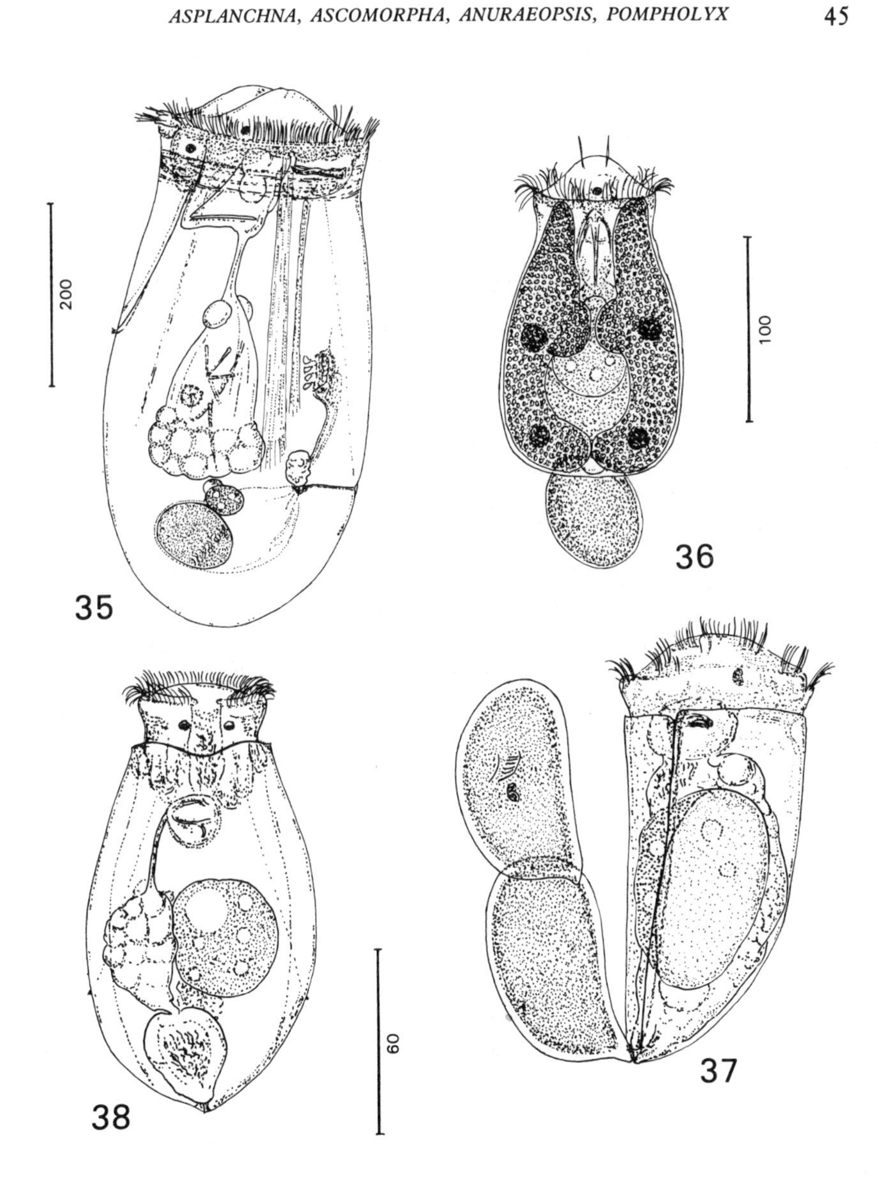

Figs 35-38. 35, *Asplanchna priodonta*. Female, lateral. 36, *Ascomorpha ecaudis*. Female, dorsal. 37, *Anuraeopsis fissa*. Female, lateral. 38, *Pompholyx sulcata*. Female, dorsal.

KEYS TO SPECIES (FEMALES)

This key attempts to include the females of all British and Irish species that have been recorded from the plankton. Under OTHER SPECIES are listed: non-planktonic members of each genus and those species found only in brackish and sea water; most European species, some of which may also occur in the British Isles but may have been overlooked; and some American and tropical species. (See 'Scope of Key', p. 5).

BRACHIONIDAE

Lorica present or absent; foot present or absent. Corona of arcs and tufts of cirri around head and cilia leading to mouth, which lies in funnel-shaped depression. (Figs 39-70). Jaws malleate (fig. 3). Planktonic, semi-planktonic and non-planktonic members.

RHINOGLENA Ehrb. 1853

One British and Irish species.

Foot very short, 2 toes. Body cone-shaped. Head with dorsal snout-like process (fig. 39*a*↓). Corona with conspicuous cirri or bristles extending in a V-shape from mouth along sides of head, with smaller cilia extending on to snout; 2 red eyes with lenses, on snout. Length <400 μm. Parthenogenetic development viviparous; resting eggs with spiny shell (fig. 39*d*). MALE similar to young female, has functional gut and trophi, length <360 μm (fig. 125)—
(Rhinops vitrea) **Rhinoglena frontalis** Ehrb. 1853

Widespread in lakes, ponds, canals; cold stenotherm, Nov.-May; sometimes abundant; planktonic.

OTHER SPECIES: *R. fertöensis* (Varga 1929) Germany, planktonic, brackish water.
TYPE SPECIES: *R. frontalis* Ehrb. 1853 (described as *Diglena frontalis).*

For a description of the genus see Stossberg (1932); for a description of the male see Hermes (1932); on *R. fertöensis* see Varga (1929, 1930).

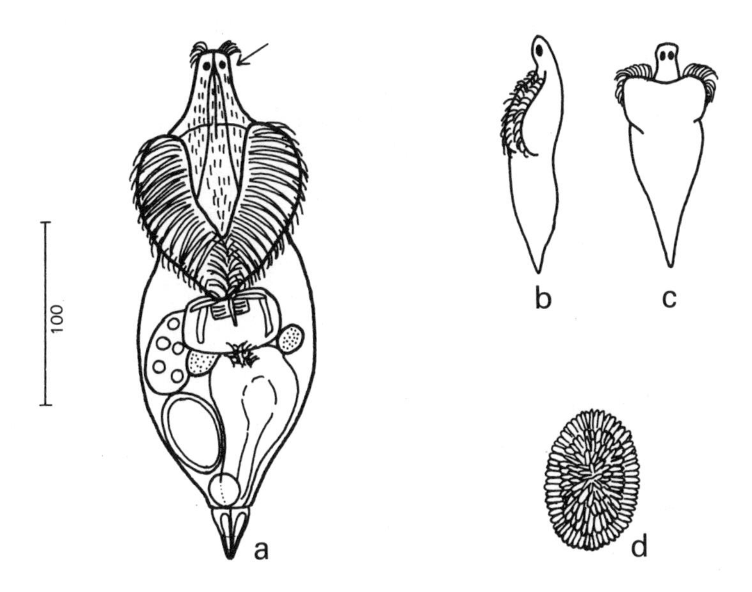

Fig. 39. *Rhinoglena frontalis*. Female, *a*, ventral, ↙ snout; *b*, lateral; *c*, dorsal; *d*, sexual egg. (*d*, after Wesenberg-Lund 1930).

EPIPHANES Ehrb. 1832

Foot long or short, 2 small toes; body cone- or sack-shaped (figs 40, 41).

1 Foot well set off from body (fig. 40). Uncus of jaw with 4 teeth. Single eye large, red. Length <600 µm. Resting eggs dark brown with granular shell. MALE with or without eye; length <210 µm; June; (fig. 137*c*)— **Epiphanes brachionus** (Ehrb. 1837) *(Notops brachionus)*

Widespread but sporadic in ponds and moor pools, in open water and between plants.

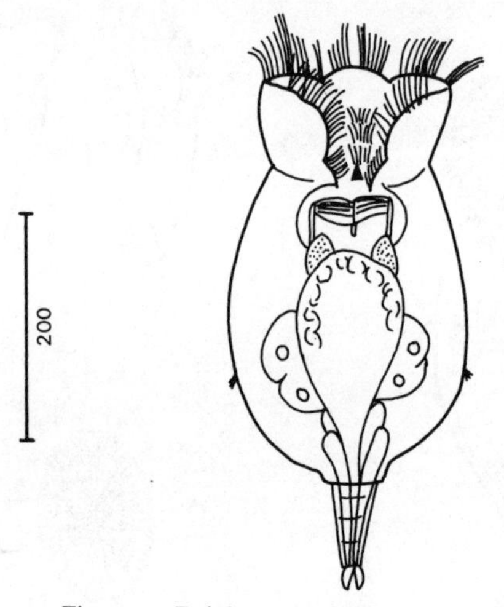

Fig. 40. *Epiphanes brachionus.* Female, ventral.

— Foot not set off from body (fig. 41). Uncus of jaw with 6 teeth (fig. 41*d*). Eyes two, colourless. Length <500 µm. Resting eggs with short bristles (fig. 41*c*). MALE <250 µm (fig. 137*a*, *b*)— *(Hydatina senta)* **E. senta** (Müller 1773)

Semi-planktonic, widespread, sometimes abundant, in eutrophic and polluted small waters, especially cattle ponds; sometimes in temporary ponds, also in large rivers and sporadic in lakes.

OTHER SPECIES: *E. clavulata* (Ehrb. 1832) (syn. *Notops clavulatus*), Britain, periphytic. *E. pelagica* (Jennings 1900) (syn. *Notops pelagicus*), N. America, Russia, planktonic. *E. macrourus* (Daday 1884) (syn. *Notops macrourus, N. mollis*), Europe, America, Africa, planktonic.

TYPE SPECIES: *E. clavulata* (Ehrb. 1832) (described as *Notommata clavulata*).

For further information on the genus see Pourriot (1965); for a description of the male see Hermes (1932).

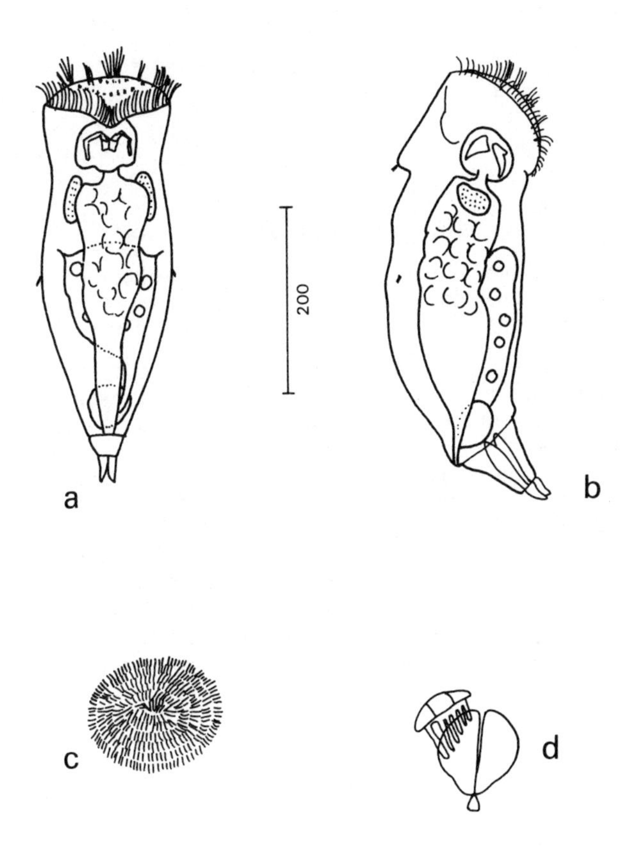

Fig. 41. *Epiphanes senta.* Female: *a*, ventral; *b*, lateral; *c*, sexual egg; *d*, jaws, uncus and manubrium of one side only. (*a*, after Weber 1898, *b*, after Hudson & Gosse 1886, *c*, after Wesenberg-Lund 1923, *d*, after Beauchamp 1909).

BRACHIONUS Pallas 1766

Foot long, very flexible, more or less wrinkled, worm-like, retractile, 2 toes; lorica usually flattened dorso-ventrally, bearing anterior spines, with or without posterior spines (figs 42-47). One red cerebral eye.

1 Anterior spines of lorica 2, median, blunt or pointed (fig. 42), sometimes small and obscured by head; no posterior spines. (Rarely spines virtually absent; or anterior spines present, flanked by 2 small points (fig. 42*d*); or 2 points on sides of lorica (fig. 42*e*)). Lorica <220 µm. Eggs carried; resting eggs brown and pitted (fig. 42*c*); MALE <105 µm (fig. 128*a-c*). Usually a perennial species, maxima spring and autumn, little polymorphism—

Brachionus angularis Gosse 1851

Cosmopolitan, planktonic in lakes, ponds and canals and brackish water.

With blunt processes on either side of foot opening (fig. 42*b*)—

var. **bidens** (Plate 1886)

— Anterior spines of lorica 4 or 6— **2**

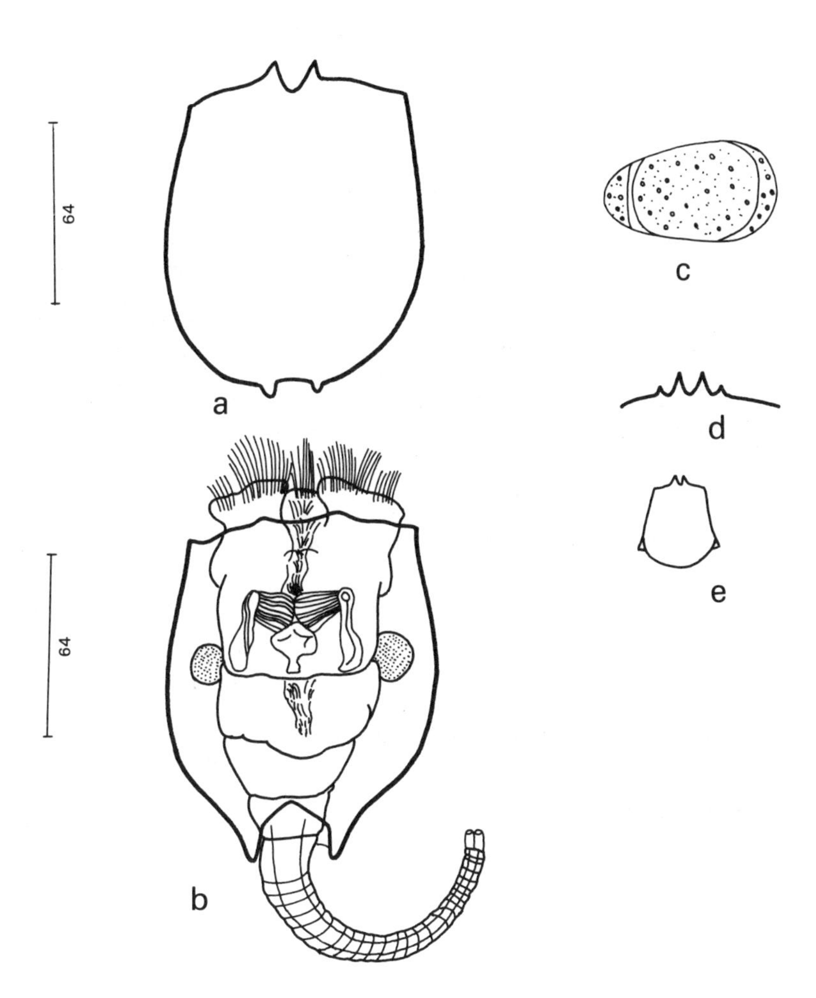

Fig. 42. *Brachionus angularis*. Female: *a*, lorica, dorsal; *b*, extended animal, ventral; *c*, sexual egg; *d*, anterior dorsal border of lorica, form with 4 points; *e*, lorica, form with lateral points. (*c*, after Voigt 1904, *d*, *e*, after Wesenberg-Lund 1930).

2 Anterior spines 4, variable but usually long, pointed, with broad bases, equal in length or median pair long (figs 1, 43). Lorica not dorso-ventrally flattened, fairly flexible; with 1 or 2 short or long posterior spines, or none; with or without spines at foot opening. Lorica <570 µm. Asexual eggs carried; resting eggs carried for a short time. MALE <120 µm, lorica slightly developed (fig. 128*g*, *h*). Perennial species, often abundant—
(B. pala) **Brachionus calyciflorus** Pallas 1766

Cosmopolitan, planktonic, in large and small waters, canals, reservoirs, large rivers, brackish water.

A very polymorphic species, many varieties and forms described. The main ones are:

Anterior spines equal in length (fig. 43*b*)— var. **pala** (Ehrb. 1838)

and with long posterior spines (<300 µm) and spines at foot opening (fig. 43*b*)— f. **amphiceros** (Ehrb. 1838)

Median anterior spines longer (fig. 43*a*), posterior spines present or absent— var. **dorcas** (Gosse 1851)

— Anterior spines 6— **3**

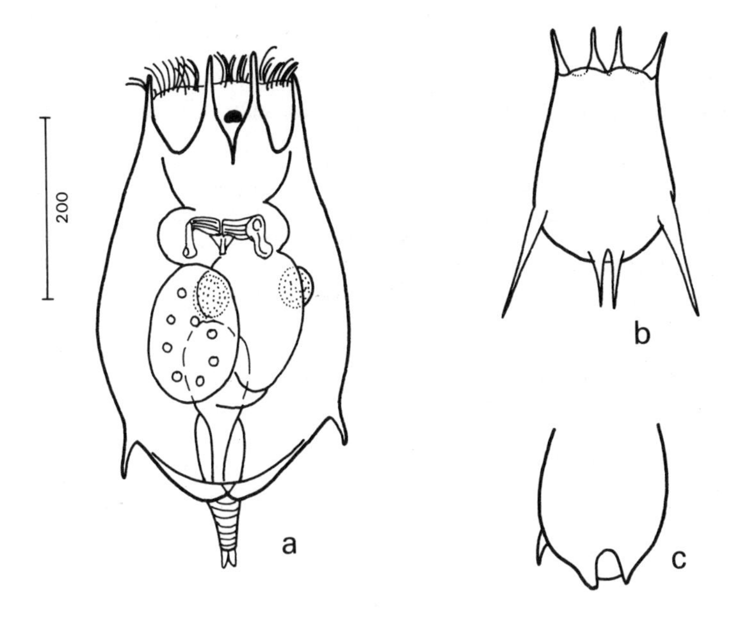

Fig. 43. *Brachionus calyciflorus.* Female: *a,* dorsal, var. *dorcas, b,* lorica, ventral, var. *pala,* f. *amphiceros; c,* posterior end lorica, ventral, with 1 spine only.

3 Foot opening tube-shaped, projecting ventrally with short lateral spines (fig. 44); posterior spines on lorica absent or present, short or long; anterior median spines usually long, curved to lateral or ventral; seasonal variation in spine length occurs. Lorica <415 µm. MALE <150 µm, May and Sept. (fig. 128*d*). Summer species with spring and autumn maxima, long-spined form summer and autumn—
Brachionus quadridentatus Hermann 1783
(B. capsuliflorus, B. bakeri)

Cosmopolitan, planktonic, also between plants; lakes, ponds, moor waters, canals and brackish and saline waters.

Posterior spines present, <$\frac{1}{3}$ lorica length (fig. 44*c*)—
var. **brevispinus** (Ehrb. 1832)

Posterior spines present, >$\frac{1}{2}$ lorica length (fig. 44*a*, *b*)—
var. **longispinosus** (Koeppel 1940)

— Foot opening not tube-shaped (figs 46, 47)— **4**

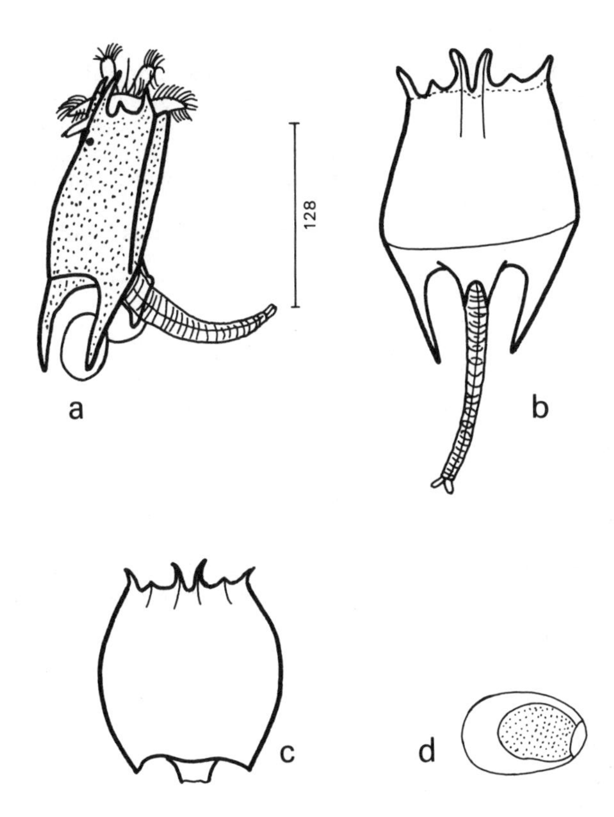

Fig. 44. *Brachionus quadridentatus.* Female: *a*, *b*, var. *longispinosus; a*, lateral, extended animal, with eggs; *b*, dorsal, head retracted; *c*, dorsal, lorica, var. *brevispinus; d*, sexual egg. (*c*, after Rousselet 1897, *d*, after Collin, Dieffenbach, Sachse & Voigt 1912).

4 Lorica with basal plate as well as ventral and dorsal plates (fig. 45*a*, *b* ↙) (see also *B. urceolaris* var. *sericus*); dorsal plate usually strongly marked with folds and net-like pattern (fig. 45*e*); no posterior spines; foot opening with 3 points. Lorica <340 µm. Asexual eggs 120×75 µm; resting eggs with short fat spines, March (fig. 45*f*), 130×90 µm. MALE <140 µm (fig. 128*f*)—

Brachionus leydigi Cohn 1862

Typically semi-planktonic in temporary pools, also in ponds, lakes and rivers, and thermal, brackish and saline waters.

A polymorphic species. The main varieties are:

Dorsal side of foot opening with 3 sharp points, ventral side with straight upper border; basal plate with 2 keels (fig. 45*a-c*)—

var. **quadratus** (Rousselet 1889)

Dorsal side of foot opening with 3 broad processes, ventral side rounded (fig. 45*d*)— var. **tridentatus** (Sernov 1901)

2 thin processes at side of foot opening; basal plate without keel (fig. 45*e*)— var. **rotundus** (Rousselet 1907)

— No basal plate (figs 46, 47*a-c*)— **5**

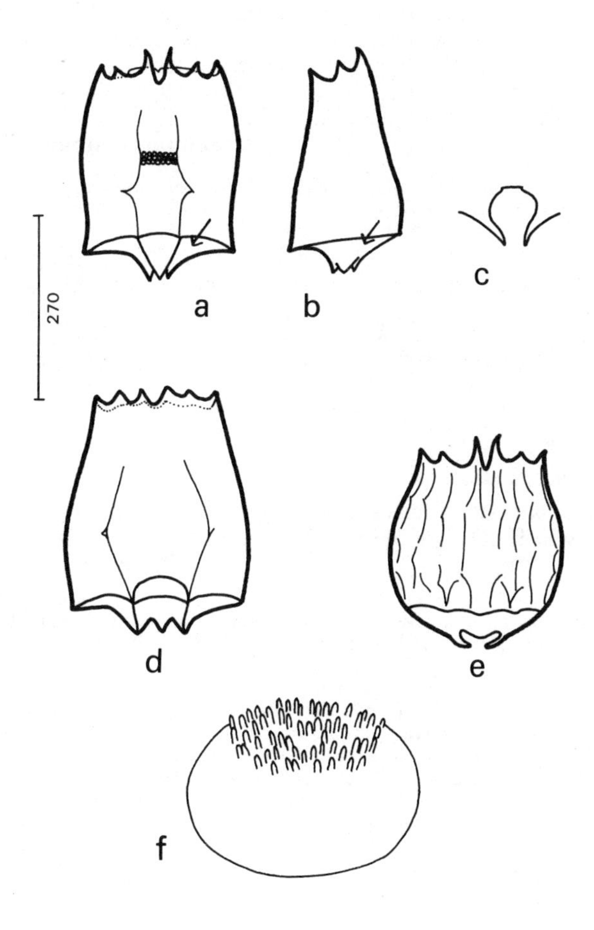

Fig. 45. *Brachionus leydigi.* Female: *a-c,* var. *quadratus; a,* lorica, dorsal; *b,* lorica, lateral, ↙ basal plate; *c,* foot opening, ventral; *d,* var. *tridentatus,* lorica, ventral; *e,* var. *rotundus,* lorica, dorsal; *f,* sexual egg. (*a-c,* after Rousselet 1889, *d,* after Voigt 1957, *e,* after Rousselet 1907, *f,* after Galliford, personal communication).

5 Each anterior spine of lorica asymmetrical, with 'shoulder' on one side (fig. 46*a* ↙); posterior spines absent; lorica very transparent, sometimes reddish. Eye more or less 4-lobed (fig. 46*a*). Lorica <280 µm. Asexual eggs carried; resting eggs yellow-brown, pitted, 140 µm long; MALE <140 µm (fig. 128*e*)—
Brachionus rubens Ehrb. 1838

Widespread; epizoic on *Daphnia, Moina* and *Polyphemus* in eutrophic ponds, also free in open water; sporadic in brackish water.

— Each anterior spine smooth and symmetrical (fig. 47*a*); posterior end rounded without spines; lorica with pitting or folds; foot opening may have spines (fig. 47*a*, *c*). Eye more or less square. Lorica <280 µm. Resting eggs brownish, pitted, <165 µm long; MALE <150 µm (figs 47*g*, 128*i*)— **B. urceolaris** Müller 1773

Cosmopolitan, in ponds, lakes, running water, moor waters, in open water and amongst plants; also in saline waters.

A variable species; also:

Dorsal surface of lorica with longitudinal striations and pitting on well set-off basal piece (fig. 47*f* ↙)— var. **sericus** Rousselet 1907
(= *B. sericus* Rousselet)

OTHER SPECIES: *B. plicatilis* Müller 1786, brackish and salt waters, Britain and Ireland. Planktonic species found in Europe, also in America: *B. falcatus* Zacharias 1898, *B. budapestinensis* Daday 1885, *B. caudatus* Barrois 1894, *B. diversicornis* Daday 1883, *B. bennini* (Leissling 1924). American and tropical planktonic or littoral species: *B. pterodinoides* Rousselet 1913, *B. bidentata* Anderson 1889 (syn. *B. furculatus*), *B. mirabilis* Daday 1897, *B. dimidiatus* (Bryce 1931), *B. nilsoni* Ahlstrøm 1940, *B. satanicus* Rousselet 1911, *B. variabilis* Hempel 1896, *B. novae-zelandiae* (Morris 1913), *B. zahniseri* Ahlstrøm 1934, *B. dolabratus* Harring 1915, *B. forficula* Wierzejski 1891, *B. havanaensis* Rousselet 1911.

TYPE SPECIES: *B. calyciflorus* Pallas 1766.

For further information on the taxonomy of the genus see Ahlstrøm (1940), Ruttner-Kolisko (1974), Voigt (1957), Gillard (1948). On seasonal variation in *B. calyciflorus* see Halbach (1970); on specificity in *B. urceolaris* and *B. quadridentatus*, Ruttner-Kolisko (1969); on egg-carrying in the genus, Sudzuki (1957a).

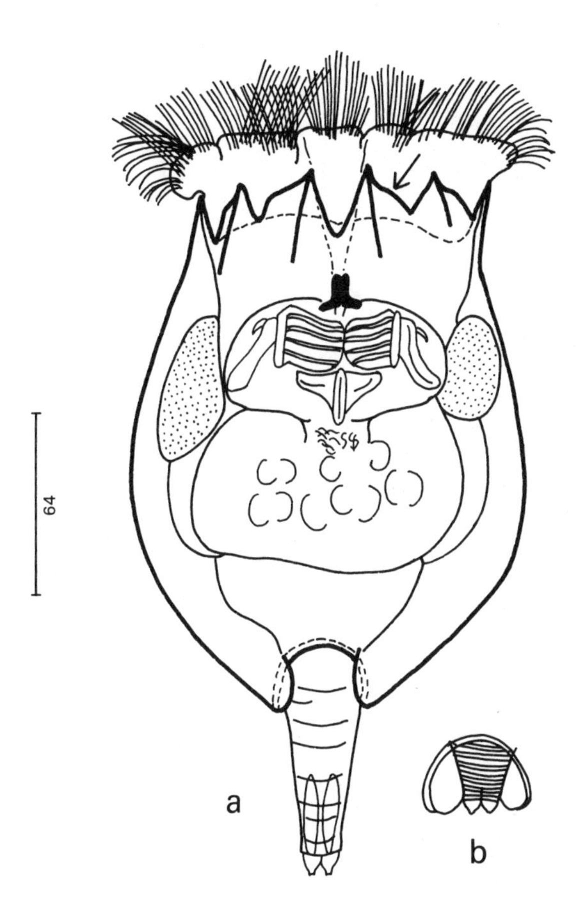

Fig. 46. *Brachionus rubens*. Female: *a*, extended, dorsal, ↙ spine; *b*, foot, retracted.

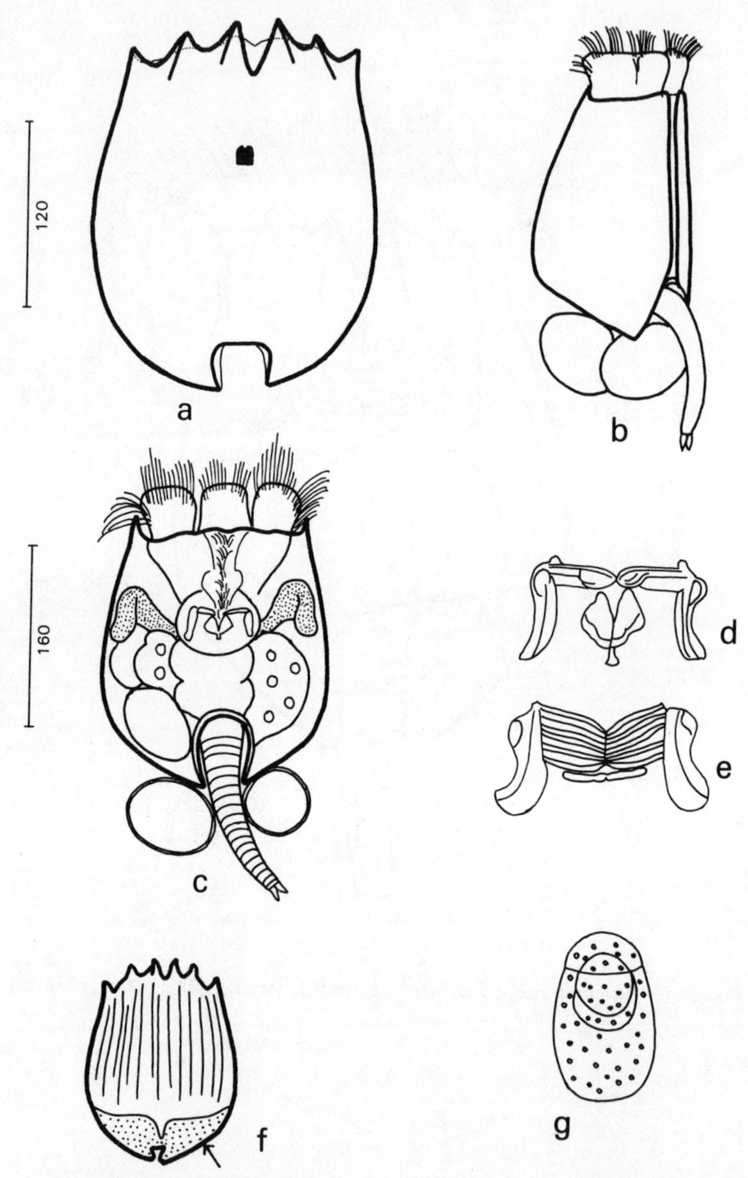

Fig. 47. *Brachionus urceolaris.* Female: *a,* lorica, dorsal; *b, c,* extended, with eggs, *b,* lateral; *c,* ventral; *d, e,* jaws, *d,* ventral; *e,* dorsal; *f,* lorica, dorsal, var. *sericus,* ↙ basal piece; *g,* sexual egg. (*f, g,* after Rousselet 1907).

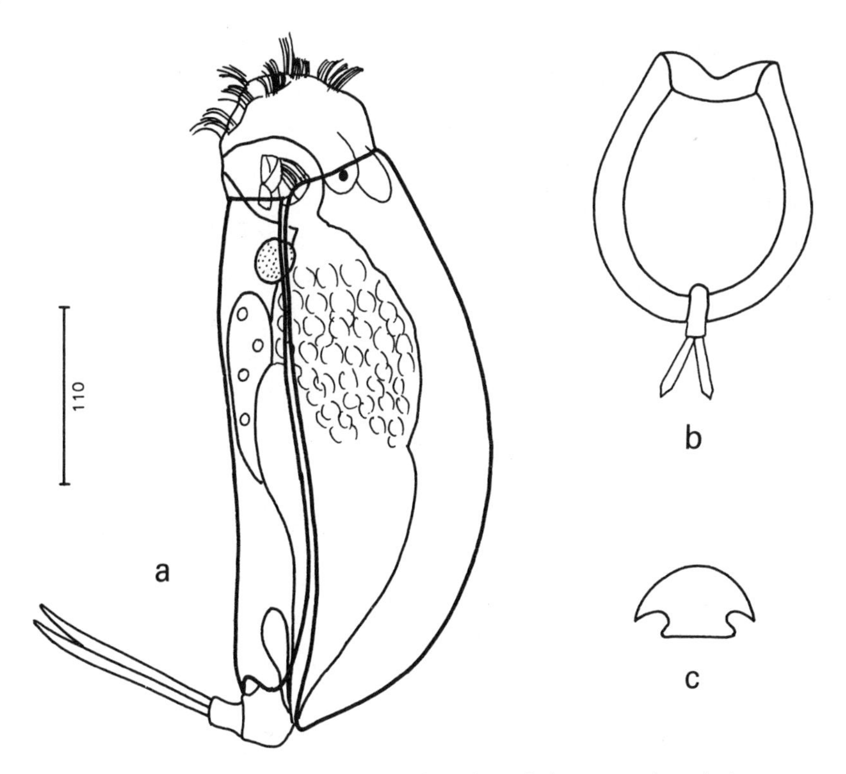

Fig. 48. *Euchlanis dilatata*. Female: *a*, lateral; *b*, lorica, ventral; *c*, lorica, cross-section. (*b*, *c*, after Harring & Myers 1930).

EUCHLANIS Ehrb. 1832

Foot of 2 sections, short, 2 toes, leaf-, sword- or spindle-shaped; lorica smooth, dorsal plate convex, with or without crest or 'wings', ventral plate flat (figs 48-50). One eye; well-developed retrocerebral organ.

1 Anterior edge of dorsal plate of lorica with gradually curving indentation (fig. 48*b*), posterior end with deep notch, dorsal plate strongly convex to roof-like in cross-section (fig. 48*c*). Uncus of jaws with 5 main and 3 accessory teeth. Length <320 µm, toes variable, <100 µm. Asexual eggs with sticky shell, laid in algae, *c*. 140×95 µm; resting eggs oval, dark brown, shell thick and granular, *c*. 110× 70 µm. MALE <260 µm (fig. 136)—

Euchlanis dilatata Ehrb. 1832

Cosmopolitan, often numerous in fresh water generally, also sporadic in brackish and saline waters. Frequently browsing amongst plants, also in plankton.

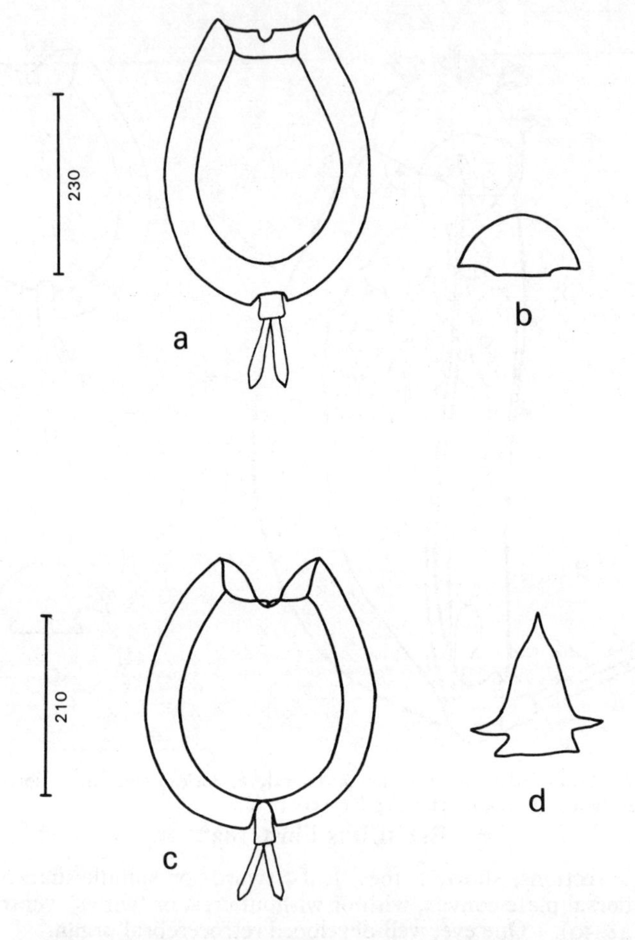

Fig. 49. *Euchlanis*. Female: *a, b, E. deflexa, a,* lorica, ventral; *b,* lorica, cross-section; *c, d, E. triquetra, c,* lorica, ventral; *d,* lorica, cross-section. (After Harring & Myers 1930).

Form with relatively shorter and finer toes, and higher dorsal plate— var. **lucksiana** Hauer 1930 (= *E. lucksiana*)

Similar to *E. dilatata,* but with high crested dorsal plate and lateral extensions or 'wings' (fig. 49*c, d*)— **Euchlanis triquetra** Ehrb. 1838

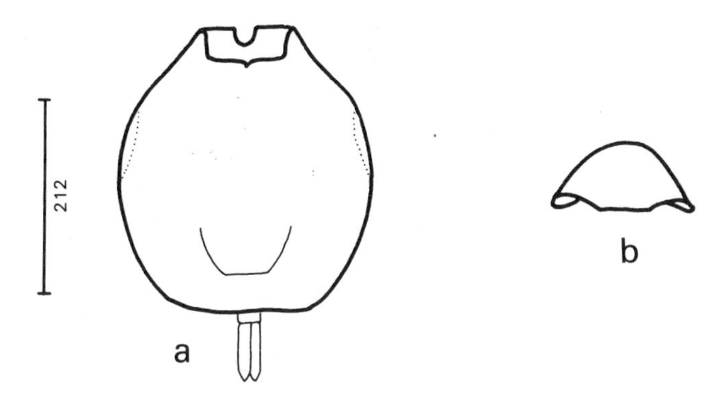

Fig. 50. *Euchlanis pyriformis*. Female: *a*, lorica, ventral; *b*, lorica, cross-section. (After Harring & Myers 1930).

— Anterior edge of dorsal plate with sharp-sided indentation (figs 49*a*, 50*a*), posterior end rounded. Uncus of jaws with 4 main and 2 accessory teeth— **2**

2 Lorica almost circular in outline, lateral edges of dorsal plate partly turned under (fig. 50). Toes almost parallel-sided. Length <320 µm, toes 85 µm— **E. pyriformis** Gosse 1851

Widespread but sporadic in lakes, ponds and rivers, usually between plants, also in open water.

— Lorica oval in outline, lateral edges of dorsal plate not turned under, dorsal plate somewhat convex (fig. 49*a*, *b*). Length <340 µm, toes 90 µm— **E. deflexa** Gosse 1851

Widespread in fresh water generally, including rivers, usually between plants, sometimes in plankton.

OTHER SPECIES, British and Irish, found among plants: *E. parva* Rousselet 1892 (syn. *E. oropha* according to Harring) — similar to *E. dilatata; E. lyra* Hudson 1886 — similar to *E. deflexa; E. meneta* Myers 1930 (syn. *E. oropha* according to Lucks); *E. proxima* Myers 1930; *E. incisa* Carlin 1939.

TYPE SPECIES: *E. dilatata* Ehrb. 1832.

For further information on the taxonomy of the genus see Harring & Myers (1930), Voigt (1957); for an account of variability and hybridization in the genus see Parise (1963), Ruttner-Kolisko (1974); also on the genus see Bülow (1954), Carlin (1939, 1943), Parise (1966).

ANURAEOPSIS Lauterborn 1900

One British and Irish species.

Foot absent; lorica smooth, egg-cup shaped, dorsal and ventral plates joined down each side by fold of elastic membrane, no spines (fig. 51). Single large red eye. Lorica <120 µm. Asexual eggs tear-drop shaped, *c.* 60×30 µm, carried attached to anal appendage or egg-carrier (fig. 51*b*↙), or in a chain of 2 or 3, trailed during swimming (fig. 51*a*); resting eggs *c.* 70×40 µm, dark brown, thick-shelled and ornamented (fig. 51*c*), July-Oct.; MALE <90 µm (fig. 133)—
Anuraeopsis fissa (Gosse 1851)
(Anuraea fissa, A. hypelasma, Anuraeopsis hypelasma.)

Planktonic in ponds, canals, bog pools; sporadic in lakes and inland saline waters.

OTHER SPECIES: *A. navicula* (Rousselet 1910), America, Java, Sumatra, planktonic.
TYPE SPECIES: *A. fissa* (Gosse 1851) (described as *A. hypelasma).*
For a revision of the genus see Bērziņš (1962); see also Gillard (1948); for egg-carrying see Sudzuki (1957b).

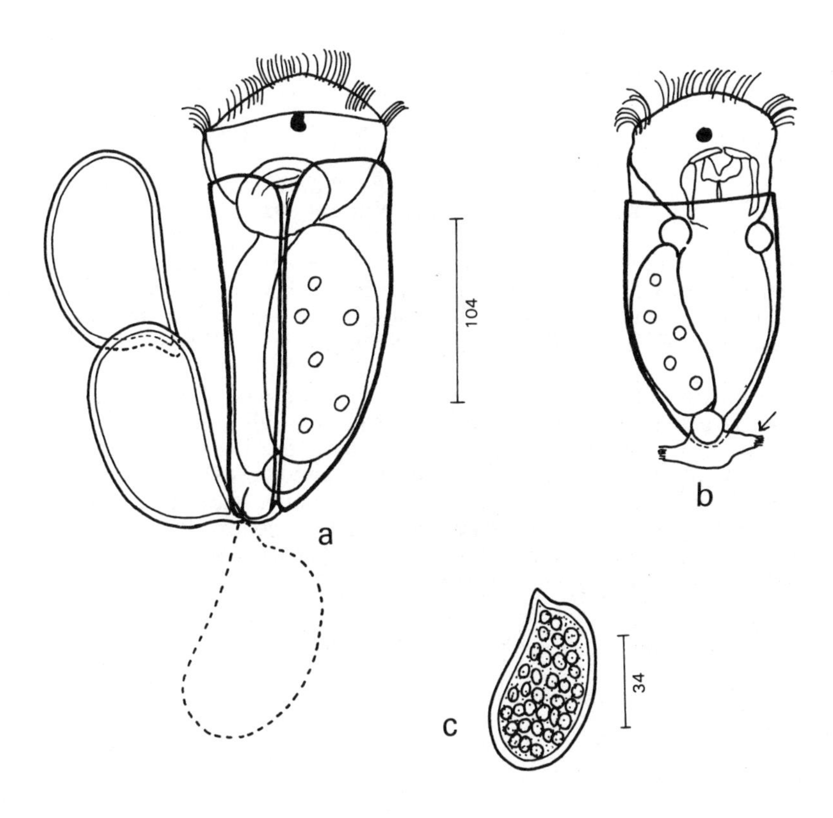

Fig. 51. *Anuraeopsis fissa.* Female: *a*, with eggs, lateral, broken line shows trailing position of eggs during swimming; *b*, ventral, showing egg carrier extended ↙; *c*, sexual egg.

KERATELLA Bory de St Vincent 1822

No foot; lorica dorsally curved, ventrally flattened or concave, with 6 anterior spines of almost equal length, with or without posterior spines; dorsal surface usually with pattern of lines (often seen well in specimens caught in surface film). (Figs 52-59). Single red eye. Eggs carried.

1 Dorsal plate of lorica with central line, on either side of which is a net-like pattern of lines and 'fields' enclosed or partially enclosed by them (figs 52*a*, 53); median single posterior spine (figs 52, 53), which may be long to short, often varying seasonally and cyclically in length, or may even be absent. Lorica <320 μm. Resting eggs smooth, brown or dark-coloured, opaque, Sept.-Nov. MALE <90 μm, lorica with 3-part dorsal plate (fig. 131*a*, *b*)— **Keratella cochlearis** (Gosse 1851) *(Anuraea cochlearis, K. stipitata)*

Often very numerous, usually perennial with two maxima a year; cosmopolitan, planktonic, in large and small waters, canals, also brackish and saline water.

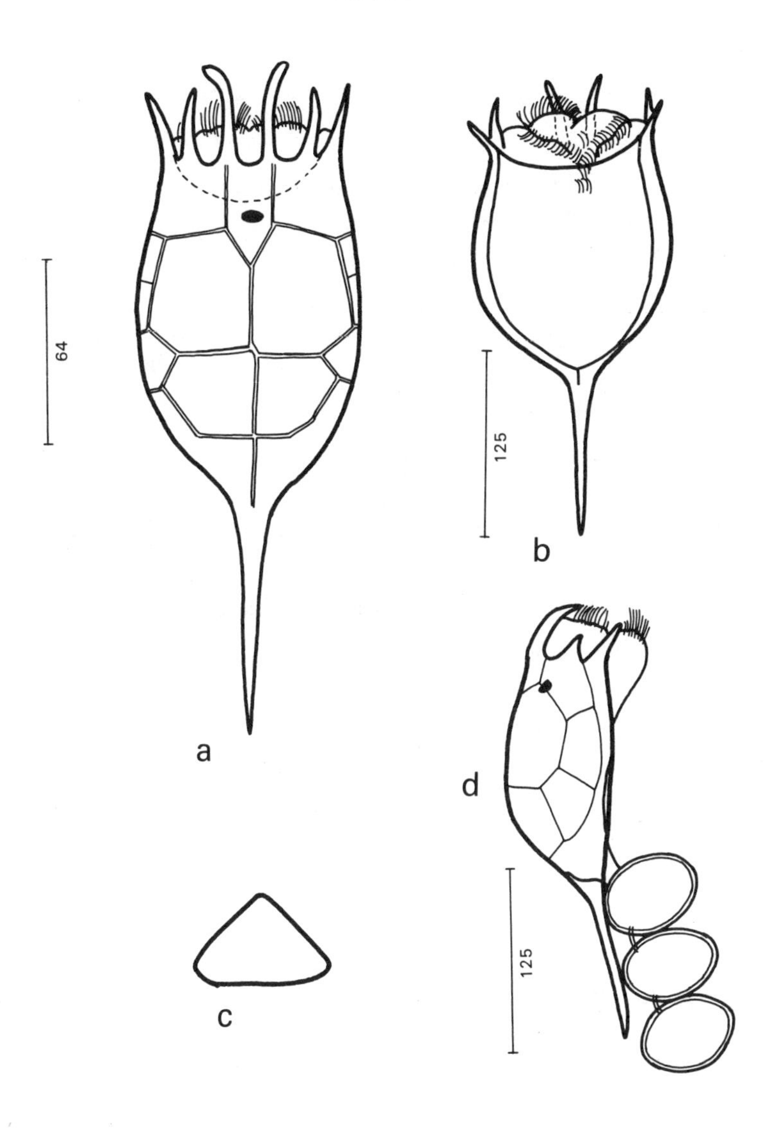

Fig. 52. *Keratella cochlearis.* Female: *a,* dorsal; *b,* ventral; *c,* lorica, cross-section; *d,* with eggs, lateral.

Keratella cochlearis (contd.)

Very polymorphic. Many forms have been described and can be arranged in groups or series, thus:

Central line of lorica pattern straight or with only a slight 'kink'; pattern symmetrical (fig. 53*d*)— **tecta** series
Includes forms with long posterior spine (fig. 53*a*) (f. *macracantha*) through medium-spined forms (f. *typica*) to short-spined (f. *leptacantha*, f. *tuberculata*) to spineless (f. *tecta*). (In this series also f. *regularis*, f. *faluta*, f. *punctata*).

Long-spined forms occur in colder, oligotrophic waters; spineless forms in warmer, eutrophic waters.

Central line of lorica pattern with definite kink; pattern asymmetrical (fig. 53*c*)— **irregularis** series
Includes long-spined forms (f. *connectens*), medium-spined (f. *irregularis*, f. *angulifera*) to spineless (f. *ecauda*); (also f. *mixta*).

In warmer waters, especially in the tropics.

Stout lorica covered with small teeth or spines, pattern symmetrical or asymmetrical (fig. 53*b*)— **hispida** series
Includes long-spined (f. *robusta*), medium (f. *hispida*) to short-spined (f. *micracantha*); (also f. *taurocephala*, f. *crassa*).

In small, shallow waters.
(For correlation of form with environmental factors see Ruttner-Kolisko (1974), especially p. 43, fig. 28).

—(1) Dorsal lorica with central row of fields, from which lines extend to either side (figs 54-59). Lateral posterior spines one, two or absent (figs 54-59)— **2**

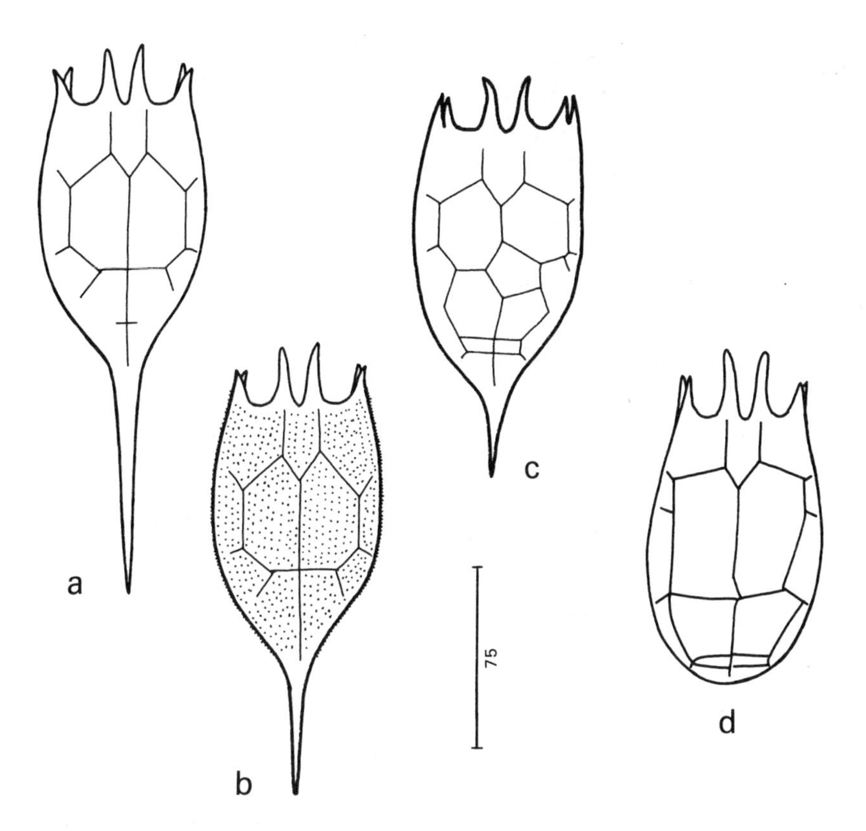

Fig. 53. *Keratella cochlearis.* Female: lorica, dorsal; *a,* f. *macracantha; b,* f. *hispida; c,* f. *irregularis; d,* f. *tecta.* (*a, b,* after Pejler 1962, *c,* after Ahlstrøm 1943 and Pejler 1962).

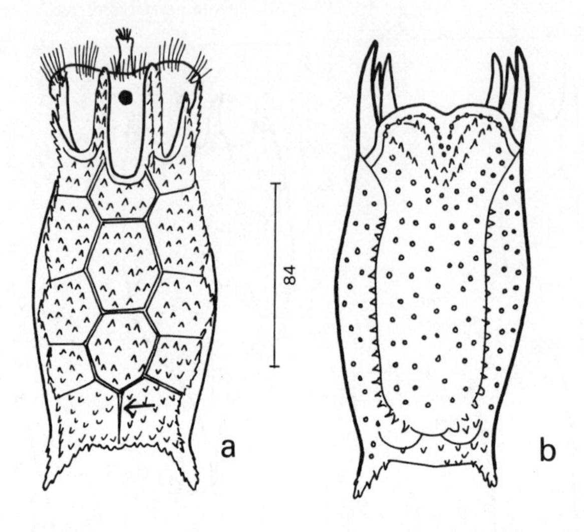

Fig. 54. *Keratella serrulata.* Female: *a,* extended, dorsal, ↙ posterior median line; *b,* lorica, ventral.

2 Central row of fields terminates posteriorly in a median line (fig. 54*a* ↙); lorica covered with short spinelets (fig. 54); posterior spines equal or unequal, or may be very short, present only as pronounced corners, or absent. Lorica <300 µm. Egg constricted into 2 unequal parts; resting eggs have yellow-brown shell with small tubercles, dark contents, 90×65 µm—

Keratella serrulata (Ehrb. 1838)
probably = *K. falculata* (Ehrb. 1838)

Widespread, sporadic but may be numerous; mostly in moor and acid waters between *Sphagnum* but also in lakes and ponds in open water.

Form with single asymmetric posterior spine—
var. **levanderi** Lie-Pettersen 1909

— Central row of fields terminates posteriorly in 2 lateral lines (figs 55-59)— **3**

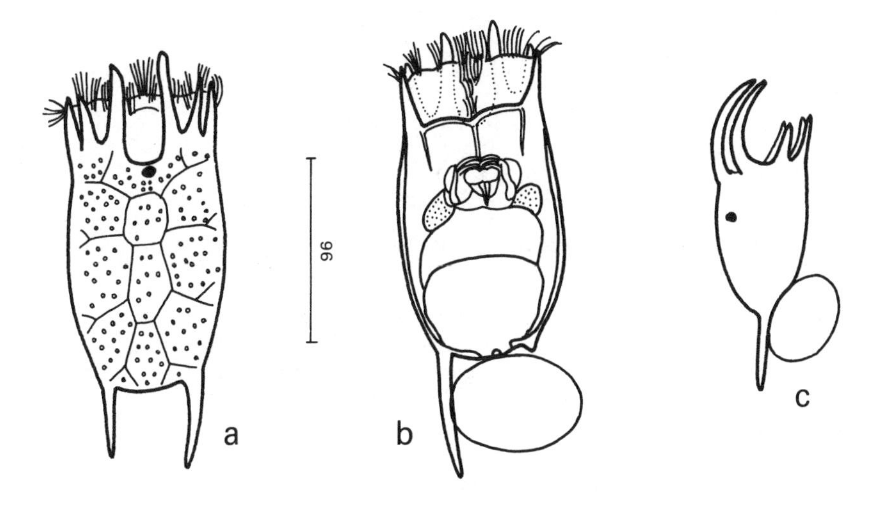

Fig. 55. *Keratella valga.* Female: *a*, extended, dorsal; *b*, ventral, with egg; *c*, lorica, lateral, with egg.

3 Lorica broader at anterior end (fig. 55*a*); posterior spines usually unequal in length, or only 1 present, or absent; some populations show little variation in spine length, with R. spine $\frac{1}{3}$-$\frac{1}{2}$ body length, but in others all stages of reduction of L. spine may be observed. Lorica <*c.* 250 µm. Resting eggs with prickly shell; MALE <90 µm—

K. valga (Ehrb. 1834)
Possibly a form of *K. quadrata*

In ponds, temporary pools, canals; planktonic.

— Lorica not broader at anterior end (figs 56-59)— **4**

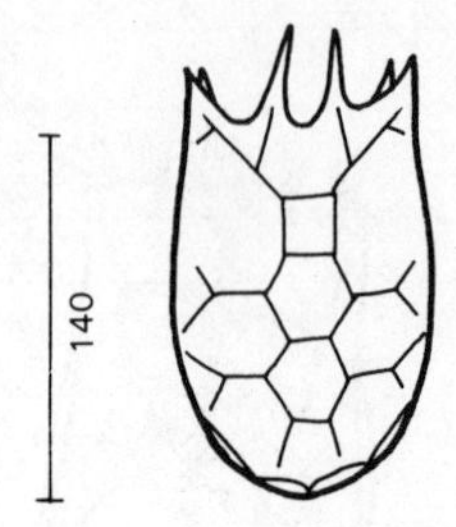

Fig. 56. *Keratella ticinensis*. Female: lorica, dorsal. (After Carlin 1943).

4 Dorsal lorica with 3 completely enclosed fields in central row (fig. 56); posterior border of lorica with row of small marginal fields (fig. 56); no posterior spines. Resting eggs with short spines which may be curved at ends, 76×54 µm— **Keratella ticinensis** (Callerio 1920)

In plankton of ponds.

— Dorsal lorica with only 2 completely enclosed fields in central row (figs 57-59)— **5**

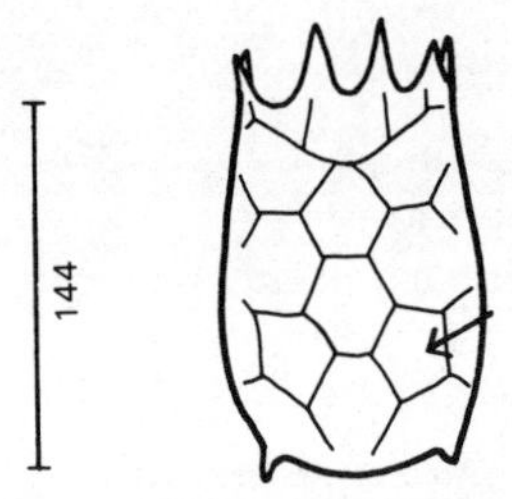

Fig. 57. *Keratella testudo*. Female: lorica, dorsal, ↙ enclosed lateral field. (After Carlin 1943).

5 Lateral fields 2, one on each side of central row, completely enclosed (fig. 57↙); lorica thick-set, less than twice as long as broad; posterior spines short, strong, rarely only one or none. Lorica $<$180 µm. Resting eggs with strong folds; MALE $<$84 µm, April— **K. testudo** (Ehrb. 1832)

In plankton of ponds.

— Lateral fields not completely enclosed (figs 58, 59); lorica about twice as long as broad— **6**

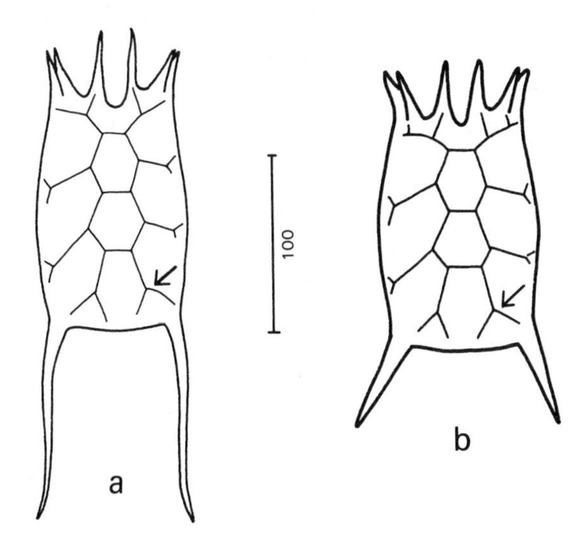

Fig. 58. *Keratella quadrata.* Female: lorica, dorsal, *a,* long-spined form; *b,* shorter-spined form; ↙ forked posterior lines. (*a,* after Carlin 1943).

6 Anterior enclosed median field of lorica more or less hexagonal (fig. 58); median row of fields terminates posteriorly in forked lines (fig. 58↙); posterior spines usually long to medium, thin, sometimes short, usually of equal length, parallel or diverging, sometimes only one spine or none. Lorica <350 μm. Resting eggs with many folds; male eggs often carried in a chain. MALE <100 μm (fig. 131*c, d*)—
K. quadrata (Müller 1786)

Widespread throughout fresh waters of all types; also saline waters.

Very polymorphic. Forms include:

Delicate forms with long posterior spines; lorica <350 μm—
f. **frenzeli**
= f. *divergens*

In large lakes with lower even temperatures.

Thick lorica with strong granulations— f. **reticulata**
= f. *dispersa.*

Probably pond forms.

See also *K. valga* (p. 71).

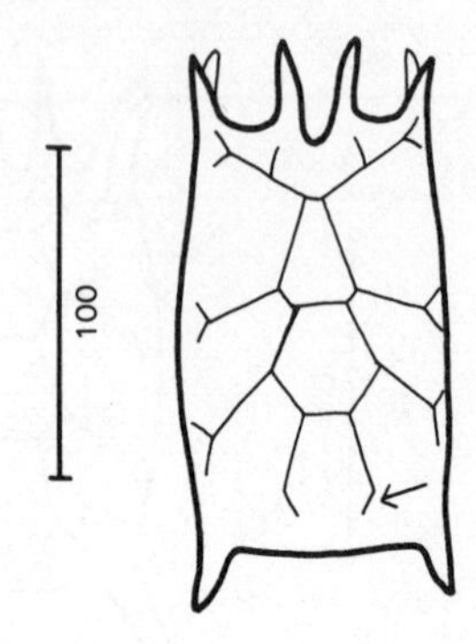

Fig. 59. *Keratella hiemalis.* Female: lorica, dorsal, ↙ unforked posterior lines. (After Carlin 1943).

— Anterior enclosed median field more or less triangular (fig. 59); median row of fields terminates posteriorly in unforked lines (fig. 59↙); posterior spines of medium length, parallel or slightly diverging. Little variation in form; acyclic. Lorica <300 µm. Males unknown— **Keratella hiemalis** Carlin 1943

Perennial species in oligotrophic lakes, cold stenotherm, forms pseudo-sexual eggs in rising temperatures.

OTHER SPECIES: *K. cruciformis* (Thompson) 1892, planktonic, marine and brackish waters of North and Baltic Seas & Atlantic Ocean. Tropical planktonic spp; *K. americana* Carlin 1943 (syn. *K. gracilenta*), *K. tropica* (Apstein 1907), *K. procurva* (Thorpe 1891), *K. lenzi* Hauer 1953).

TYPE SPECIES: *K. quadrata* (Müller 1786) (described as *Brachionus quadratus).*

For revisions of taxonomy of the genus and descriptions of variability in form see especially Ahlstrøm (1943), Bērziņś (1954a, 1955), Voigt (1957); also Amrèn (1964), Bērziņś (1954b), Carlin (1943, 1945), Fergg (1964), Förster (1951), Gillard (1948), Hutchinson (1967), Klement (1955), Pejler (1957, 1962), Rühmann (1954), Ruttner-Kolisko (1949); on egg-carrying see Sudzuki (1957a).

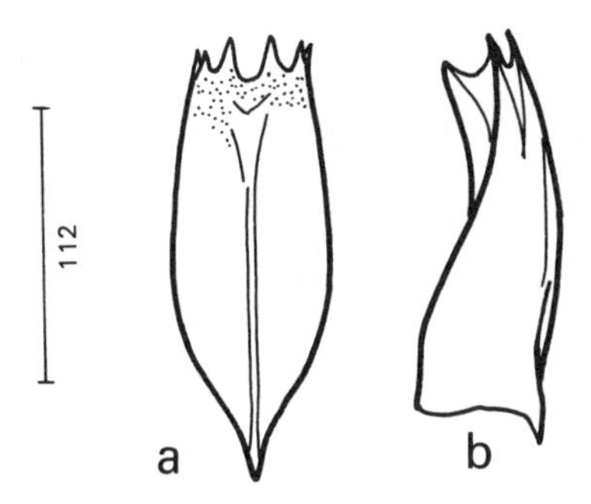

Fig. 60. *Argonotholca foliacea.* Female: *a*, lorica, dorsal; *b*, lorica, lateral. (After Olofsson 1917).

NOTHOLCA Gosse 1886 and ARGONOTHOLCA Gillard 1948

No foot; lorica usually shield-shaped, dorsally curved, with 6 anterior spines of almost equal length, with or without posterior spines or processes; dorsal surface usually with longitudinal striations. (Figs 60-63). Single eye, red or purple.

1 Dorsal surface of lorica with median ridge prolonged into pointed spine at posterior end (fig. 60). Ventral surface protruding at posterior end (fig. 60*b*). Lorica <180 µm. Resting eggs and males unknown— **Argonotholca foliacea** (Ehrb. 1838) *(Anuraea foliacea, Notholca foliacea)*

Lakes and ponds, usually sporadic, rather numerous in winter, in open water and between plants. Also in brackish water. Cosmopolitan.

— No ridge down lorica— **2**

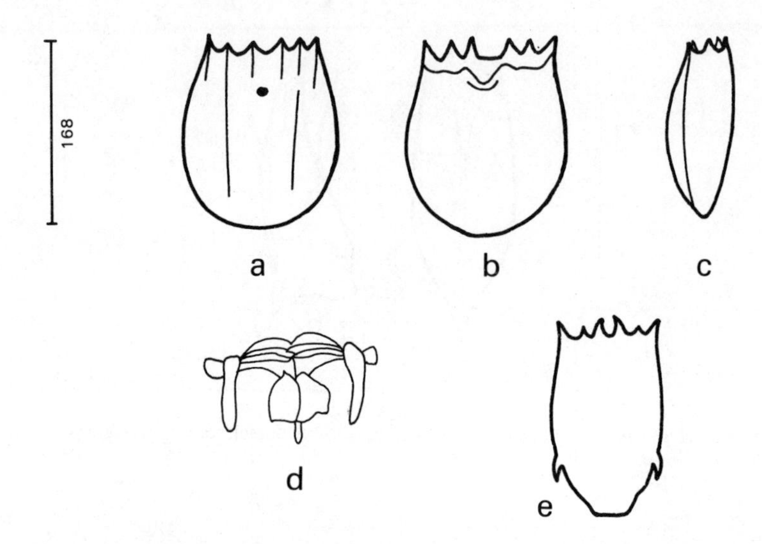

Fig. 61. *Notholca.* Female: *a-d, N. squamula; a,* lorica, dorsal; *b,* lorica, ventral; *c,* lorica, lateral; *d,* jaws; *e, N. striata,* lorica, dorsal. (*e,* after Gillard 1948).

2 Lorica not much longer than broad (L : B <2); posterior end rounded, never with spine or process (fig. 61*a, b*); strongly flattened dorso-ventrally; anterior spines short. Lorica <180 µm. Resting eggs yellow-brown— **Notholca squamula** (Müller 1786)

Fresh water only. (Considerable confusion exists in the literature and in the records over certain species, in particular *N. squamula* and *N. striata* (Müller 1786) (syn. *N. bipalium*). *N. striata* occurs only in brackish, sea or inland saline waters, while *N. squamula* is confined to fresh water. *N. striata* is further distinguished from *N. squamula* by the presence of 2 small lateral spines on the lorica (fig. 61*e*) though these may be difficult to see.) Planktonic. Lakes, ponds, canals, winter and spring.

— Lorica about twice as long as broad, or longer— **3**

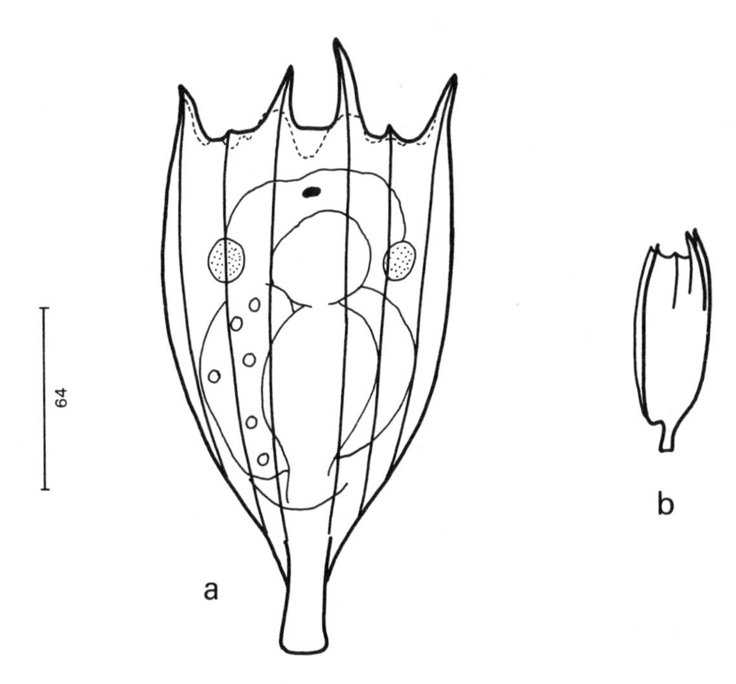

Fig. 62. *Notholca labis.* Female: *a*, dorsal, retracted; *b*, lorica, lateral.

3 Posterior end of lorica with broad-ended process, somewhat set off from lorica (fig. 62); process variable but wider at distal end; lorica about twice as long as broad, dorso-ventrally flattened (L : B = 2), <250 µm— **N. labis** Gosse 1887

Widespread, sporadic in lakes and ponds and brackish water, planktonic.

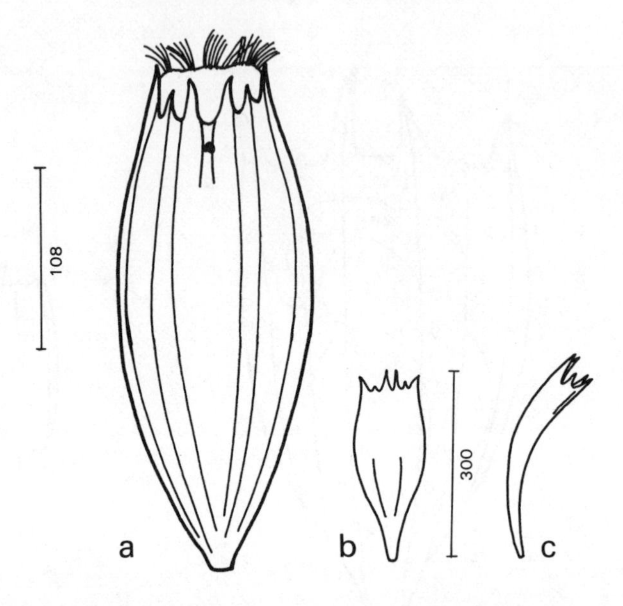

Fig. 63. *Notholca acuminata.* Female: *a*, extended, dorsal, posterior process short; *b*, *c*, lorica, posterior process long; *b*, dorsal; *c*, lateral.

— Posterior end of lorica rounded or extending gradually into a more or less well-developed process, usually not set off (fig. 63); process variable but wider at proximal end; lorica about 2-3 times as long as broad (L : B >2), <300 µm. Resting eggs larger than asexual eggs, dark coloured; MALE in April— **Notholca acuminata** (Ehrb. 1832)

Widespread in lakes and ponds and brackish water in the cold season, planktonic.

OTHER SPECIES: *N. cinetura* Skorikov 1914, a very large form from Russia and Sweden. *N. limnetica* (Levander 1901), not very clearly described; see Carlin (1943). *N. cornuta* (Carlin 1943), Sweden; see Carlin (1943). *N. psammarina* (Buchholz 1956), psammophile. *N. lapponica* Ruttner-Kolisko 1966, psammophile. *N. frigida* Jaschnow 1922, Scandinavia and L. Baikal. Also other species from L. Baikal.

TYPE SPECIES: *Notholca: N. striata* (Müller 1786) (described as *Brachionus striatus).* The original genus description for *Notholca* Gosse is in Hudson & Gosse (1886). *Argonotholca: A. foliacea* (Ehrb. 1838) (described as *Anuraea foliacea*).

For further information on taxonomy in both genera see Carlin (1943), Ruttner-Kolisko (1974); also Amrèn (1964), Buchholz and Rühmann (1956), Focke (1961), Gillard (1948), Pejler (1957, 1962).

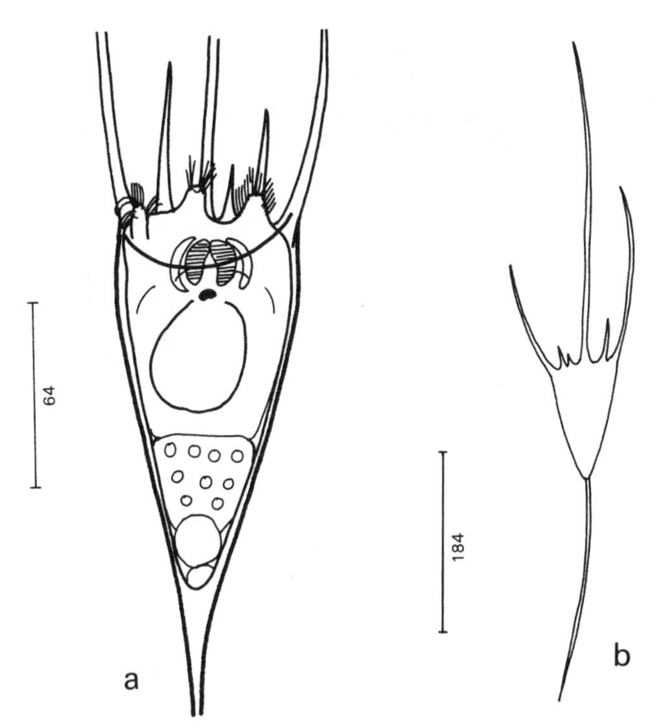

Fig. 64. *Kellicottia longispina.* Female: *a*, corona extended, ventral, bases only of long spines shown; *b*, lorica, dorsal.

KELLICOTTIA Ahlstrøm 1938

One British and Irish species.

Foot absent; lorica smooth, 3-sided, anteriorly bearing 6 spines of different lengths, median spine as long as or longer than body; lorica tapering posteriorly to a single spine (fig. 64). Lorica <860 µm. Asexual eggs carried; resting eggs thick-shelled; MALE with small spines, June— **Kellicottia longispina** (Kellicott 1879) *(Notholca longispina)*

Planktonic, limnetic, often numerous, species usually present throughout year, maxima in summer and often in winter also; winter form with longer spines. Oligotrophic lakes, also in ponds and brackish water.

OTHER SPECIES: *K. bostoniensis* (Rousselet 1908), America, planktonic.

TYPE SPECIES: *K. longispina* (Kellicott 1879) (described as *Notholca longispina).*

For further taxonomic information see Rousselet (1908), Amrèn (1964), Carlin (1943), Pejler (1962).

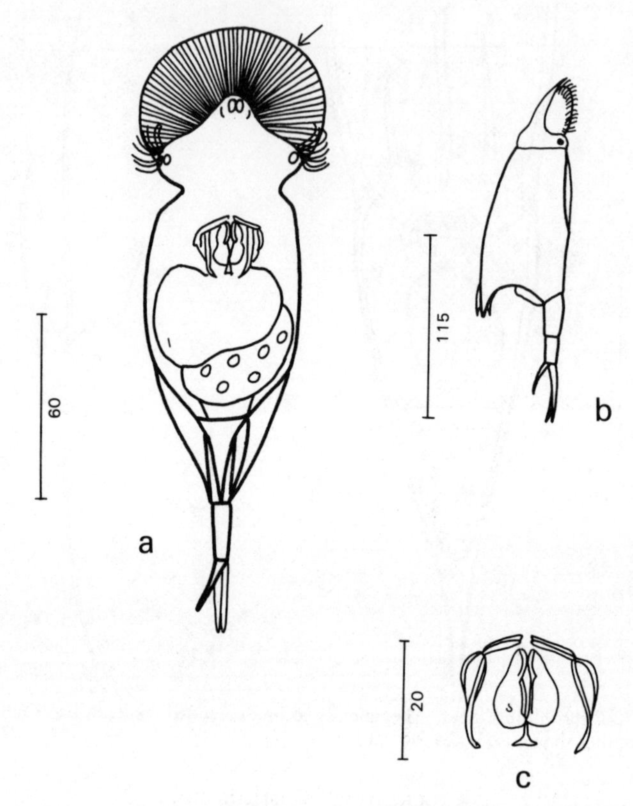

Fig. 65. *Squatinella rostrum*. Female: *a*, dorsal, ↙ head shield; *b*, lateral; *c*, jaws. (*b*, after Weber 1898).

SQUATINELLA Bory de St Vincent 1826

Foot of 2 or 3 sections, of medium length, with or without spine; 2 toes, thin and pointed; conspicuous transparent head shield over head and corona (figs 65, 66); lorica cylindrical, with 3 long or short posterior spines or projections, or spineless. Eyes 2, with lenses; jaws rather slender (fig. 65*c*). Male and resting eggs unknown.

A largely periphytic genus.

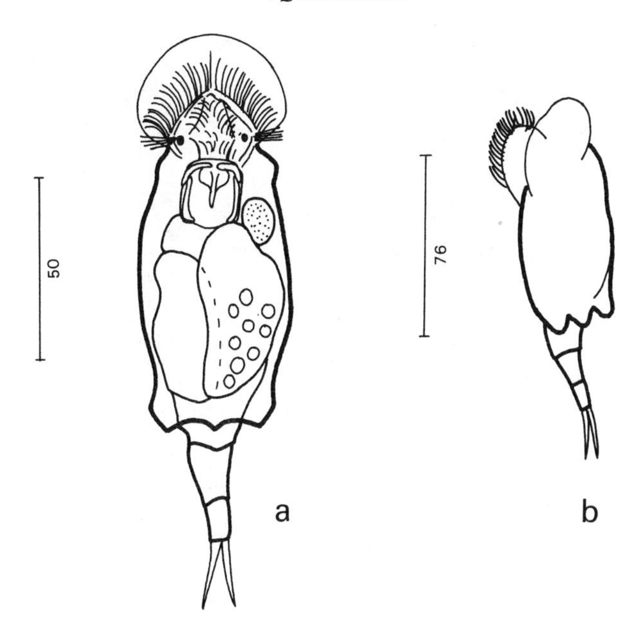

Fig. 66. *Squatinella tridentata.* Female: *a*, dorsal; *b*, dorso-lateral.

1 Foot with spine at base of toes (fig. 65*a*, *b*); posterior end of lorica with 3 short or long spines with fairly narrow bases (fig. 65). Length <220 µm, toes 28 µm— **Squatinella rostrum** (Schmarda 1846) *(Stephanops lamellaris)*

Cosmopolitan, sporadic in large and small waters, moor pools, canals, usually between plants, sometimes in open water.

— Foot without spine above base of toes (fig. 66); posterior end of lorica with 3 short or long spines with broad bases (fig. 66). Length <180 µm, toes 25 µm— **S. tridentata** (Fresenius 1858)

Widespread in ponds, moor pools and canals, usually between plants, sometimes in open water.

OTHER SPECIES: British and Irish periphytic species: *S. mutica* (Ehrb. 1832), *S. longispinata* (Tatem 1867), *S. leydigi* (Zacharias 1886), *S. bifurca* (Bolton 1884), *S. microdactyla* (Murray 1906). European species: *S. bisetata* (Ternetz 1892), *S. aurita* Wulfert 1950.

TYPE SPECIES: *S. cirrata* (Müller 1773) (described as *Brachionus cirratus).*

For further literature on taxonomy see Carlin (1939), Donner (1954), Varga (1933).

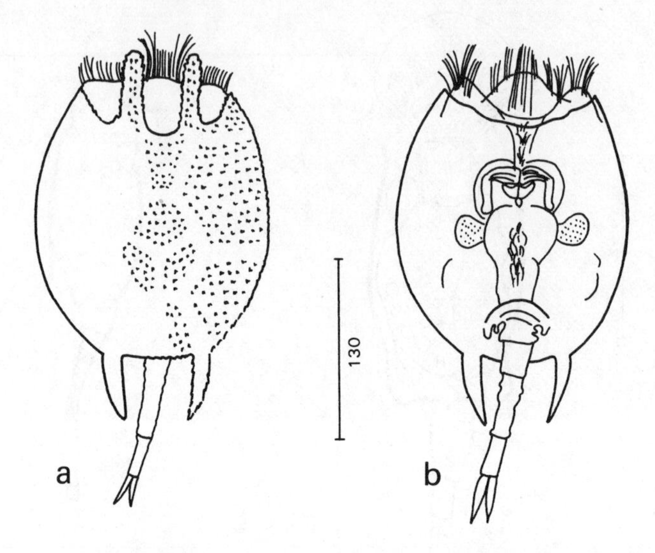

Fig. 67. *Platyias quadricornis*. Female: *a*, dorsal; *b*, ventral.

PLATYIAS Harring 1913

Foot of 3 sections, partly retractile; 2 short, pointed toes; lorica stout with 2 anterior and 2 posterior spines and covered with spinelets. Eye present or absent. (Fig. 67).
A littoral genus found amongst plants.

TYPE SPECIES: *P. quadricornis* (Ehrb. 1832) (described as *Noteus quadricornis*). Original description of genus in Harring (1913b).

For taxonomic information see Ahlstrøm (1940), Wulfert (1965).

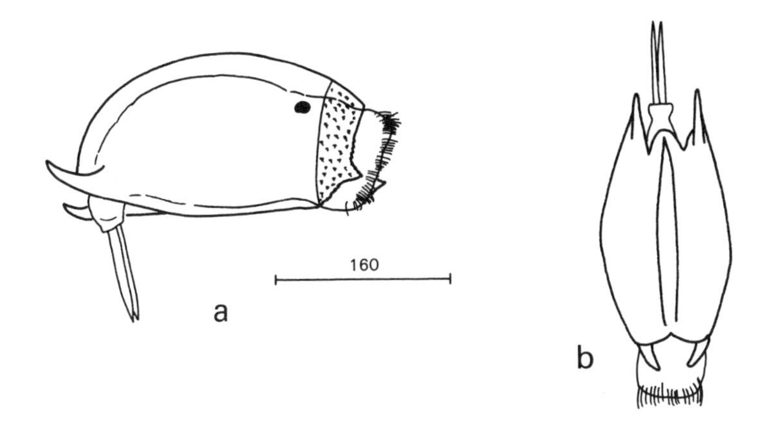

Fig. 68. *Mytilina ventralis.* Female: *a*, lateral; *b*, dorsal.

MYTILINA Bory de St Vincent 1826

Foot of 1 or more sections; 2 pointed toes, straight or curved; lorica somewhat flattened laterally, mussel-shaped in side-view, of 3 plates, with dorsal cleft and spines at posterior or all corners. One eye. (Fig. 68). Found amongst plants, littoral and benthic.

TYPE SPECIES: *M. mucronata* (Müller 1773) (described as *Brachionus mucronatus).*
For taxonomy see Rühmann (1965), Voigt (1957).

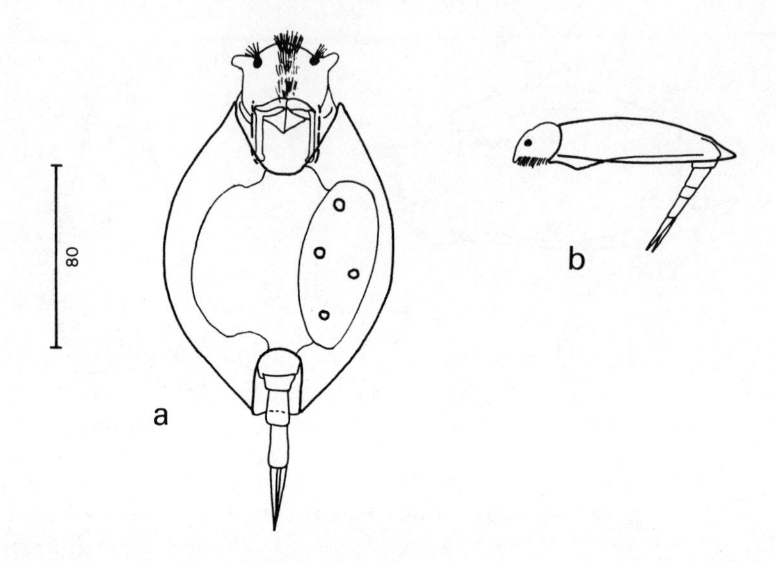

Fig. 69. *Lepadella ovalis.* Female: *a*, ventral; *b*, lateral.

LEPADELLA Bory de St Vincent 1826

Foot of 3-4 sections, proximal part concealed under lorica; foot opening deeply incised; 2 toes, short to long, pointed. Lorica one-piece, oval, egg- or pear-shaped, more or less flattened dorsoventrally, with or without dorsal crests and lateral wings; head opening narrow and semicircular. Two lateral eyes. (Fig. 69).
A benthic and littoral genus.

TYPE SPECIES: *L. patella* (Müller 1786) (described as *Brachionus patella).*
For taxonomy of genus see Voigt (1957).

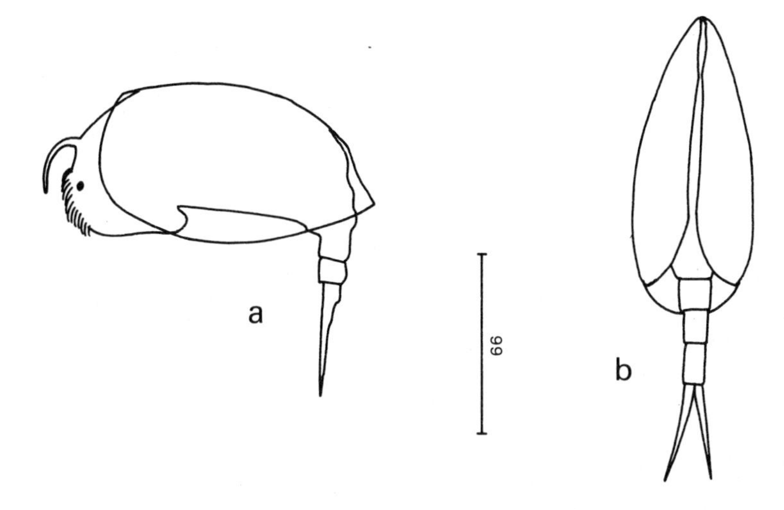

Fig. 70. *Colurella adriatica.* Female: *a*, lateral; *b*, ventral, retracted.

COLURELLA Bory de St Vincent 1824

Foot of 3-4 sections; 2 pointed toes. Lorica laterally flattened, mussel-shaped in side-view, rounded or truncated anteriorly, rounded to pointed posteriorly; small retractile head-shield. With or without 2 lateral eyes. (Fig. 70).
Benthic.

TYPE SPECIES: *C. uncinata* (Müller 1773) (described as *Brachionus uncinatus*). Original description of genus in Bory de St Vincent (1824a).

For information on taxonomy see Carlin (1943), Hauer (1924).

TRICHOCERCIDAE

Lorica present, with or without crests and spines; foot present (figs 71-79). Jaws virgate, usually asymmetrical (fig. 5*c*, *d*). Corona an anterior ring and a buccal field. Planktonic, semi-planktonic and non-planktonic members.

TRICHOCERCA Lamarck 1801

Lorica of one piece; elongated, cylindrical or spindle-shaped, or short, humped and compact; often with dorsal striated area which may be extended into 1 or 2 raised crests or keels; anterior border of lorica with or without 1 or more spines or processes; foot small, 1 or 2 sections; toes 1 or 2, equal or unequal, often with bristles at base (substyli). (Figs 71-79). Asymmetry often apparent to varying degrees in jaws, position of single eye and antennae, and size of toes. Corona often with 1 or more palps.

1 Body long, slender, cylindrical or spindle-shaped; L : B = 3:1; length <460 µm (figs 72, 74)— **2**

— Body short, sturdy, humped or curved, compact; L : B = 2:1; length <180 µm (figs 78, 79)— **6**

2 Anterior border of lorica bears hood-like projection (fig. 71); crests absent. Length <300 µm. Toes 2, unequal, crossed over each other; left toe <125 µm, right toe very short; 2 bristles (substyli). Body length : left toe >2:1; left toe : right toe >2:1. Jaws slender, more or less symmetrical. Asexual eggs carried on other rotifers, especially *Asplanchna priodonta,* and on other plankters; resting eggs brown with net-like pattern on shell; MALE <100 µm, toes 5-6 µm, late summer and winter—

Trichocerca capucina (Wierzejski 1893)

Cosmopolitan, planktonic, usually sporadic, rarely numerous; in standing waters, including moor pools; predatory on eggs carried by other rotifers, especially *Keratella* spp.

— Anterior border of lorica without hood, with 1 or 2 spines, or without processes (check in animal with retracted corona)— **3**

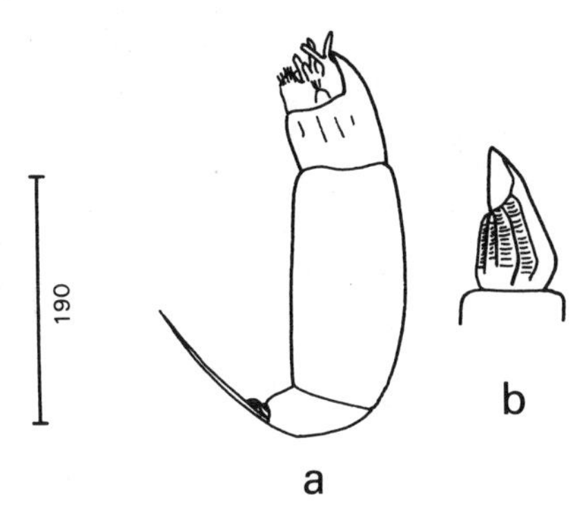

Fig. 71. *Trichocerca capucina.* Female: *a*, lateral; *b*, anterior end of lorica. (*a*, After Voigt 1957; *b*, after Jennings 1903).

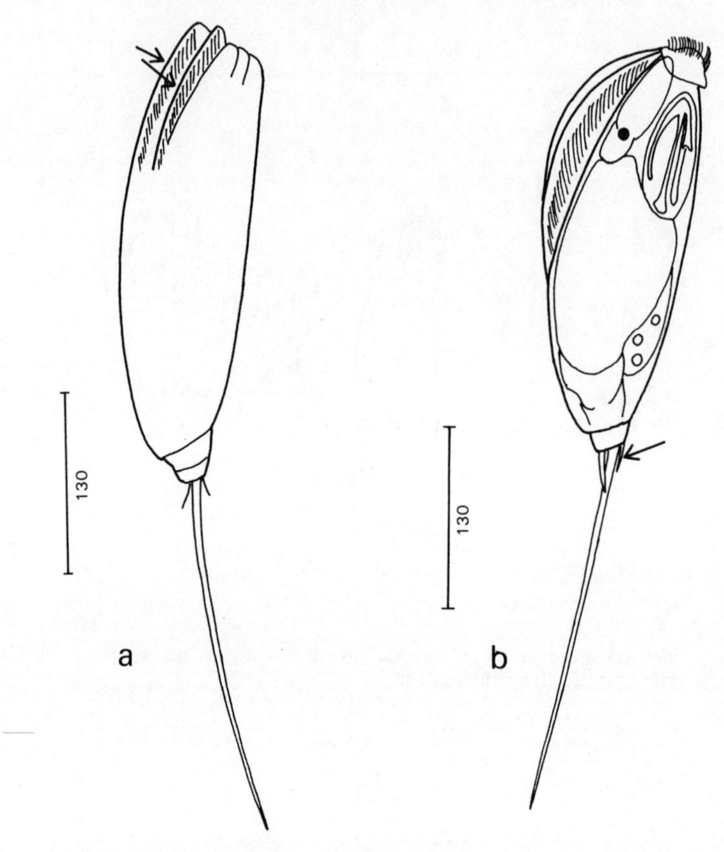

Fig. 72. *Trichocerca*. Female: *a*, *T. elongata*, lateral, ↙ crests; *b*, *T. bicristata*, lateral, ↙ substyli. (*a*, after Ruttner-Kolisko 1974).

3 Anterior border of lorica without processes; lorica usually with 2 low crests with broad furrow (fig. 72*a* ↙). Length <400 μm. Toes 2, unequal; left toe $\frac{1}{2}$-1×body length; right toe <$\frac{1}{5}$ body length, less than $\frac{1}{2}$ left toe. Bristles (substyli) present (fig. 72*b* ↙). Jaws strong. Corona with finger-like palp. Resting eggs with smooth shell, 115×55 μm— **Trichocerca elongata** (Gosse 1886)

Widespread among plants and in plankton of small and large standing waters.

Very similar to *T. elongata*, and possibly synonymous with it, but with higher crests, (fig. 72*b*), corona with 2 thin palps, length <360 μm; perennial species— **T. bicristata** (Gosse 1887)

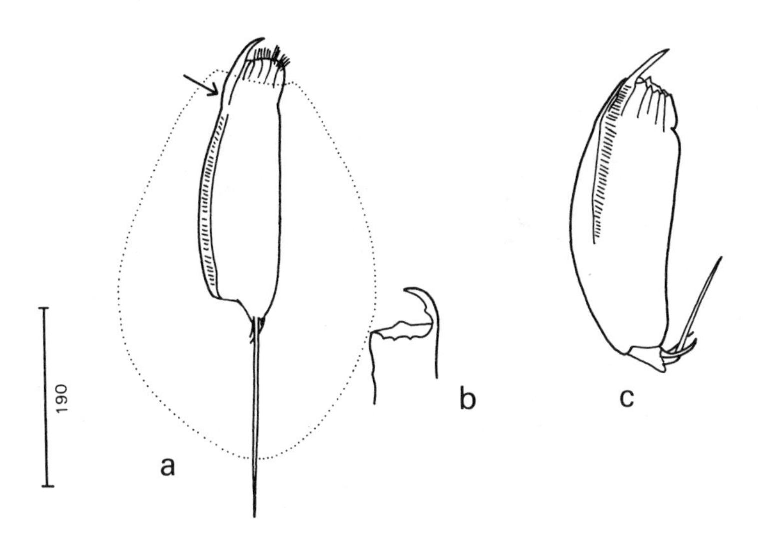

Fig. 73. *Trichocerca.* Female: *a, b, T. cylindrica; a,* lateral, in case, ↙ crest; *b,* anterior end lorica, lateral; *c, T. chattoni,* lateral. (*a,* after Lauterborn 1908, *b,* after Ruttner-Kolisko 1974, *c,* after Hauer 1938).

— Anterior border of lorica with 1 or 2 spines (figs 73-75)— **4**

4 Anterior border of lorica with 1 spine (fig. 73); lorica with striated area and with 1 crest (fig. 73*a*↙). Length <320 µm. Sometimes with gelatinous case (fig. 73*a*). Only left toe present, about as long as body. 2 bristles (substyli) present. Jaws almost symmetrical. Eggs carried; asexual eggs blue-green; MALE <80 µm (fig. 140)—
T. cylindrica (Imhof 1891)

Planktonic in eutrophic lakes and ponds; summer and winter, sporadic, rarely numerous, widespread.

Very similar to *T. cylindrica,* but right toe not completely reduced; left toe $\frac{1}{2}$ body length; body <350 µm (fig. 73*c*)—
(T. cylindrica var. *chattoni)* **T. chattoni** (de Beauchamp 1907)

Not yet recorded in British Isles.

— Anterior border of lorica with 2 spines (figs 74, 75)— **5**

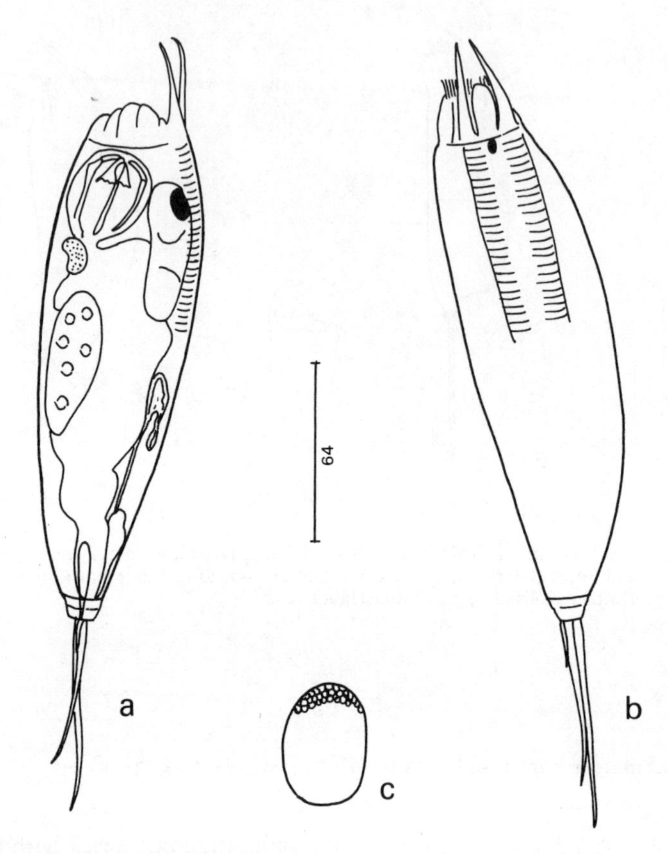

Fig. 74. *Trichocerca similis*. Female: *a*, lateral; *b*, dorsal; *c*, sexual egg. (*c*, after Wesenberg-Lund 1930).

5 Two toes, equal or almost equal (fig. 74). Crests absent, striated area present. Length <200 μm. Toes crossed over each other, *c*. $\frac{1}{3}$ body length; substyli present; foot of 1 section. Corona with 2 palps. Jaws more or less symmetrical. Eggs carried on other rotifers, especially *Brachionus angularis*— **T. similis** (Wierzejski 1893)

Planktonic, widespread in large and small waters including moor waters.

Possibly synonymous with *T. similis*, but has 2 sections in foot, 2 anterior spines displaced somewhat to right; eutrophic form, cosmopolitan— **T. birostris** (Minkiwicz 1900)

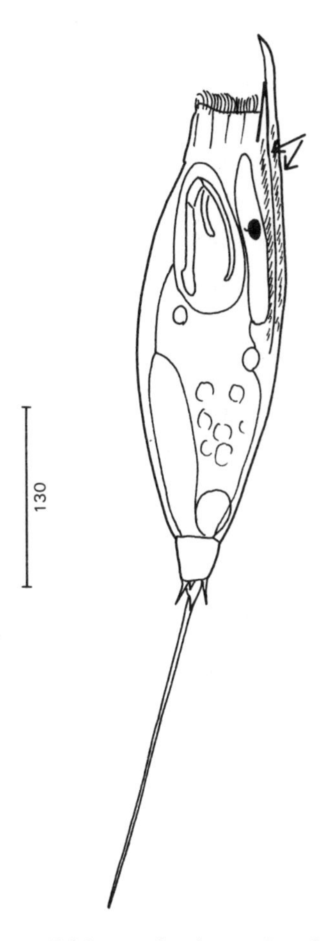

Fig. 75. *Trichocerca longiseta.* Female, lateral, ↙ crests.

— Left toe equal to $\frac{1}{2}$ body length or more; right toe very short or absent; substyli present. Lorica with striated area and 2 low crests (fig. 75 ↙); 2 anterior spines unequal; length <320 μm—

T. longiseta (Schrank 1802)

Cosmopolitan, planktonic, also between plants and in sand; probably cold stenotherm.

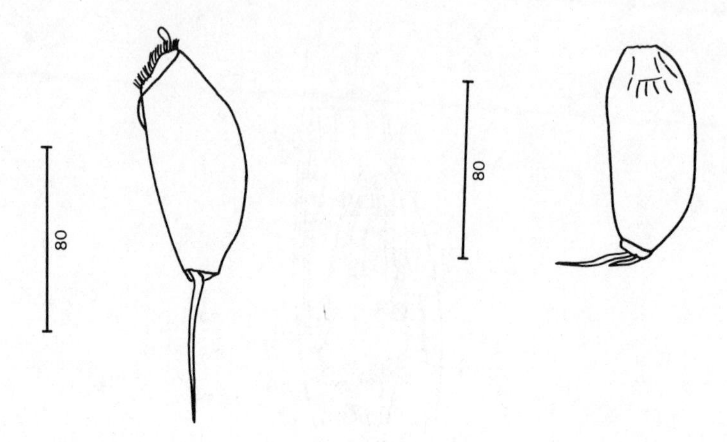

Fig. 76. *Trichocerca pusilla.* Female, lateral. (After Jennings 1903).

Fig. 77. *Trichocerca inermis.* Female, lateral. (After Voigt 1957).

6(1) One toe only, or 2 extremely unequal toes (fig. 76); left toe equal to $\frac{1}{2}$ body length or more; right toe very short or absent, left toe with characteristic bend; 1 or 2 bristles (substyli) present. Crests and striated area absent; lorica with smooth border (with folds in contracted animal). Length <115 µm. Corona with palp. Asexual eggs and male eggs attached to other rotifers, especially *Brachionus angularis*; MALE <60 µm, August— **Trichocerca pusilla** (Jennings, 1903) (probably syn. *T. stylata* (Gosse 1851))

Cosmopolitan, planktonic, usually sporadic, June-Sept.; lakes and ponds, moor waters and brackish water.

— 2 toes, equal or rather unequal (figs 77, 78)— **7**

7 Left toe 3× length of right toe (fig. 77); left toe *c.* $\frac{1}{3}$ body length. Length <135 µm. Crests absent; anterior border of lorica with undulations— **T. inermis** (Linder 1904) *(Diurella dixon-nuttalli)*

Sporadic in lakes.

— Toes equal or almost equal (fig. 78)— **8**

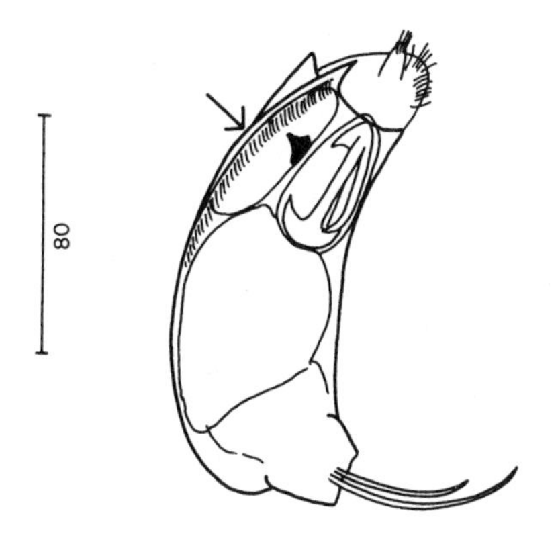

Fig. 78. *Trichocerca porcellus.* Female, lateral, ↙ crest.

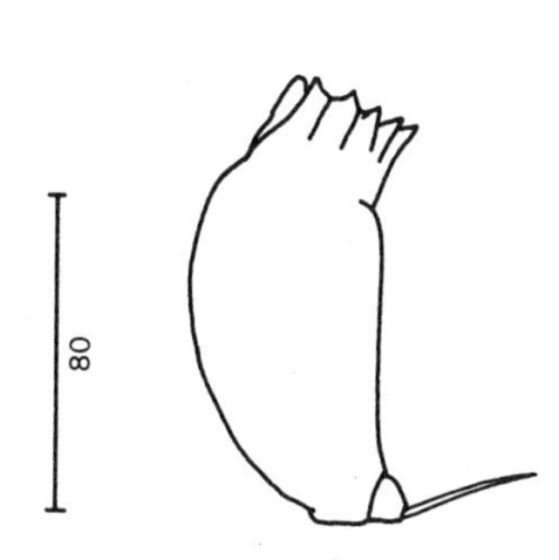

Fig. 79. *Trichocerca rousseleti.* Female, lateral. (After Jennings 1903).

8 Crest present (fig. 78 ↙); striated area present; anterior border of lorica with 2 dorsal spines. Length <180 µm. Toes *c.* $\frac{1}{3}$ body length; left toe <60 µm, right toe <45 µm; bristles (substyli) present. Corona with palp. Jaws very strongly asymmetrical— *(D. tigris)* **T. porcellus** (Gosse 1886)

Cosmopolitan; predominantly littoral; also in plankton, when eggs attached to other plankters, especially *Melosira, Fragilaria* and *Dinobryon;* large and small waters, large rivers; perennial species, often numerous.

— Crest absent (fig. 79); striated area present; anterior border of lorica with 8 or 9 teeth. Length <115 µm, body more than 2× length of toes; left toe <30 µm. Corona with long palp. Asexual eggs attached to *Melosira*— **T. rousseleti** (Voigt 1901)

Usually sporadic in plankton of lakes, also ponds; cold stenotherm?

OTHER SPECIES: planktonic, not recorded from Britain and Ireland: *T. rosea* Stenroos 1898; *T. ruttneri* Donner 1953; *T. marina* Daday 1890, a brackish water species.

TYPE SPECIES: *T. rattus* (Müller 1776) (described as *Trichoda rattus*).

For the non-planktonic species, see Voigt (1957). For further information on the planktonic species see Bērziņš (1960), Carlin (1943), Donner (1953), Myers (1934b), Nipkow (1961), Pourriot (1970).

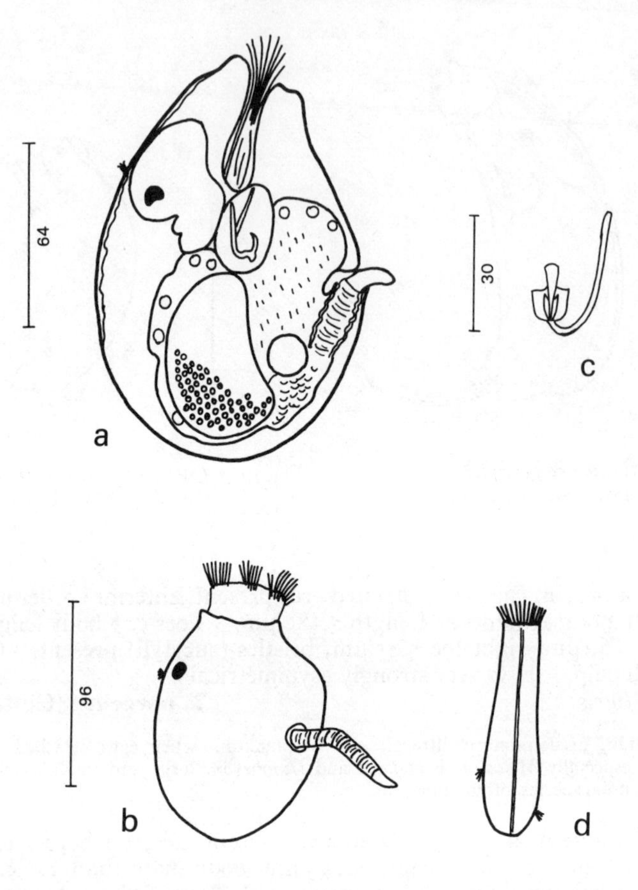

Fig. 80. *Gastropus stylifer*. Female: *a*, corona and foot retracted, lateral; *b*, corona and foot extended, lateral; *c*, jaws; *d*, dorsal. (*c*, after Voigt 1957).

GASTROPODIDAE

Lorica present or absent; foot present or absent (figs 80-85). Jaws virgate (fig. 5). Corona a simple ring. Stomach with blind extensions. A planktonic family.

GASTROPUS Imhof 1888

Foot opening ventral, foot appearing somewhat wrinkled, 1 or 2 toes; lorica thin, delicate, flask-shaped, laterally flattened (figs 80-82). One large red cerebral eye. Eggs dropped or laid.

1 One toe (fig. 80); foot partly wrinkled. Body often coloured, stomach blue and green, body cavity pink. Jaws lie reversed. Length <240 μm. Resting eggs with spines; MALE <80 μm, August (fig. 141*c*, *d*)— **Gastropus stylifer** Imhof 1891

Summer form or perennial, sporadic but may be numerous; planktonic, widespread in lakes, ponds, and moor waters.

— 2 toes— **2**

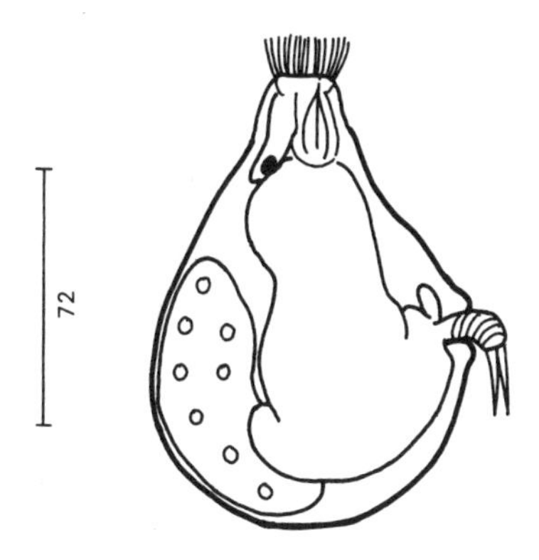

Fig. 81. *Gastropus minor*. Female, lateral. (Partly after Wesenberg-Lund 1930).

2 Anterior end narrowed (fig. 81); foot slightly wrinkled, toes long and pointed. May be coloured yellowish or brownish. Length <130 μm, toes <23 μm. Slow swimmer. Eggs laid on algae— **G. minor** (Rousselet 1892)

Planktonic, widespread but usually sporadic in large and small waters (also amongst plants).

— Anterior end not narrowed (fig. 82); lorica moderately stiff; foot short, toes pointed. Stomach contents usually yellowish. Length <360 µm. Fast swimmer. MALE <130 µm, March (fig. 141*a*, *b*)—
Gastropus hyptopus (Ehrb. 1838)

Planktonic, usually sporadic in lakes, ponds, ditches, canals, moor waters.

TYPE SPECIES: *G. stylifer* Imhof 1891.

For further information on taxonomy see Carlin (1943), Wesenberg-Lund (1930).

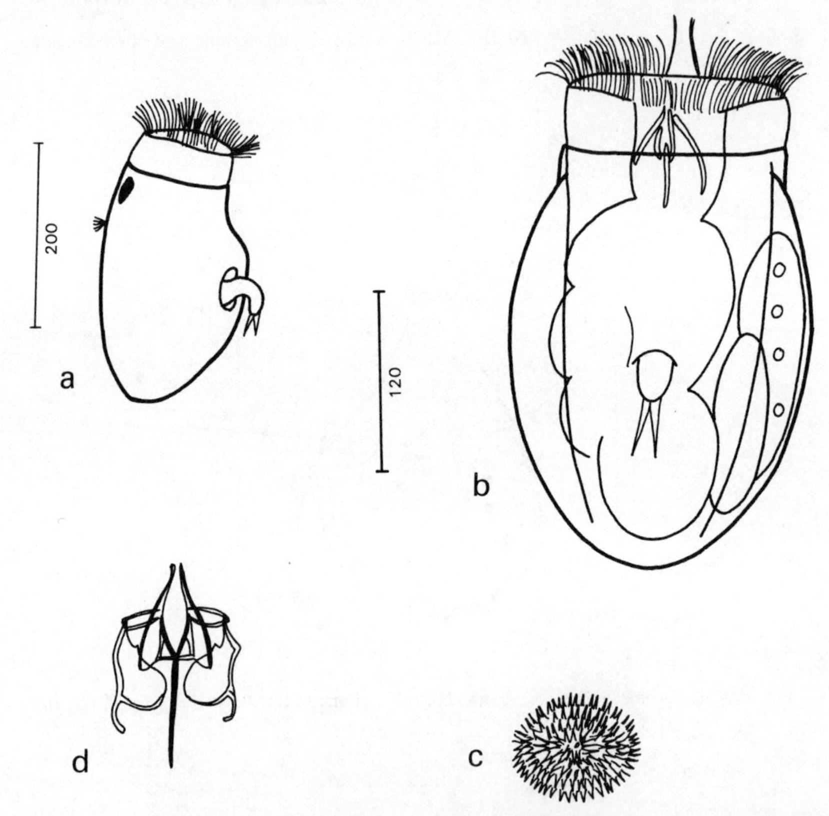

Fig. 82. *Gastropus hyptopus.* Female: *a*, extended, lateral; *b*, ventral; *c*, sexual egg; *d*, jaws. (*c*, after Wesenberg-Lund 1930, *d*, after Wulfert 1939).

ASCOMORPHA Perty 1850

No foot; body sack-like, oval or somewhat flattened dorso-ventrally; cuticle thin or very slightly stiffened. (Figs 83-85). Stomach with extensions or lobes, filling most of body cavity; stomach wall with algal cells or chromatophores. One red cerebral eye.

1 With palp on head (figs 84, 85)— **2**

— Without palp. Body sack-shaped, cross-section circular, cuticle thin (fig. 83). Conspicuous because of extensive stomach filled with dark greenish or brownish contents, zoochlorellae in stomach wall (fig. 83*a*); 4 small dark masses (accretion bodies) which are not expelled, arranged as fig. 83*c* (see also fig. 83*a*↙). Sometimes with gelatinous case. Length <200 µm. MALE <90 µm (fig. 143*a*)—
(Sacculus viridis) **Ascomorpha ecaudis** Perty 1850

Planktonic, cosmopolitan and perennial species, with maximum in summer; in lakes and ponds, canals and moor waters (also in saline waters).

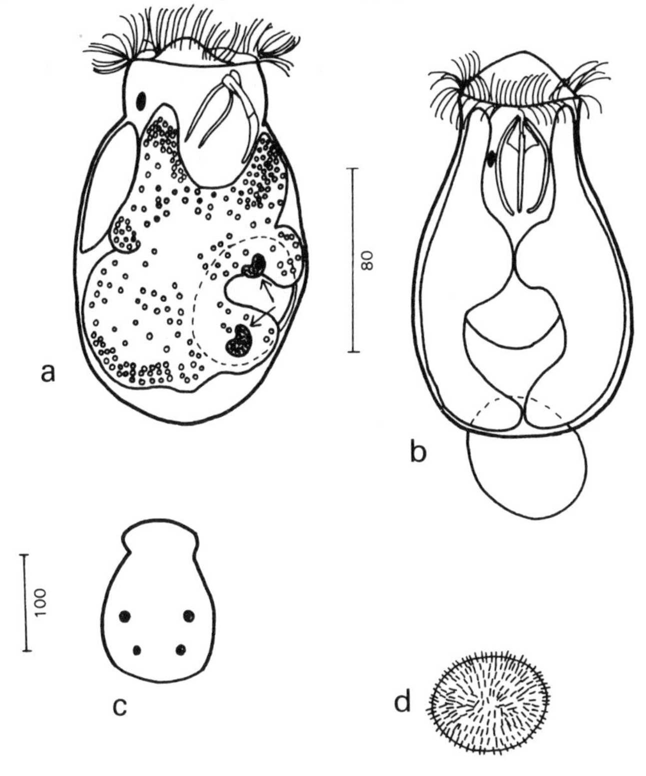

Fig. 83. *Ascomorpha ecaudis.* Female: lateral, showing accretion bodies ↙ ; *b*, dorsal, showing gut extensions; *c*, dorsal, showing arrangement of accretion bodies; *d*, sexual egg. (*c*, after Carlin 1945, *d*, after Wesenberg-Lund 1930).

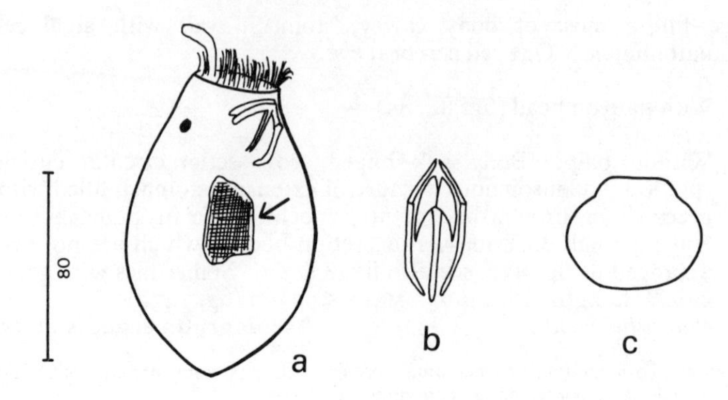

Fig. 84. *Ascomorpha saltans*. Female: *a*, lateral, showing single accretion body ↙ ; *b*, jaws, ventral; *c*, cross-section of body.

2 Palp finger-like, usually curved to dorsal (fig. 84). Body oval, cross-section as fig. 84*c*; cuticle slightly stiffened. Body contains 1 dark irregular mass (accretion body), which is not expelled (fig. 84*a*↙); chromatophores in stomach wall. Length <140 μm. Swims rather jerkily (wiggles!)— **Ascomorpha saltans** Bartsch 1850
(*A. hyalina*, *A. agilis* in part)

Widespread, planktonic in lakes, ponds, canals, moor waters (also sporadic in shore sand).

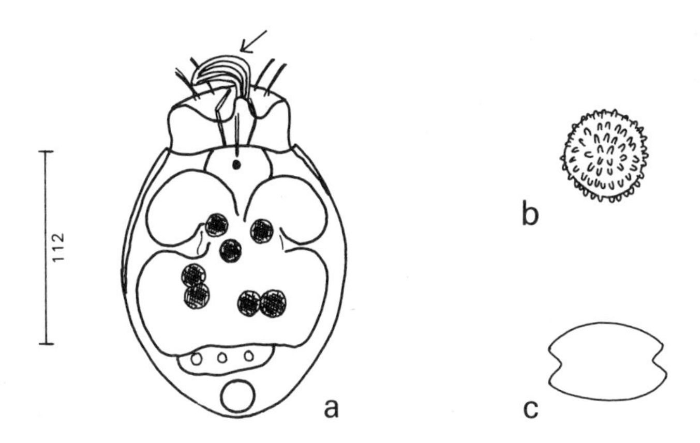

Fig. 85. *Ascomorpha ovalis.* Female: *a*, dorsal, ↙ palp; *b*, sexual egg; *c*, lorica, cross-section. (*a*, *b*, after Ruttner-Kolisko 1938, *c*, after Weber 1898).

— Palp (fig. 85*a*↙) sickle-shaped, flattened from side to side. Body more or less flattened dorso-ventrally, cross-section as fig. 85*c*; cuticle somewhat stiffened to form a very thin lorica. Numerous accretion bodies (fig. 85*a*); chromatophores in stomach wall. Length <200 µm— **A. ovalis** Carlin 1943
(Anapus ovalis, A. testudo, Ascomorpha testudo, Chromogaster testudo, C. ovalis)

Planktonic, widespread but sporadic in lakes and ponds, rivers (also saline waters); usually summer form but perennial in some localities; occurs with *Ceratium* on which it feeds.

OTHER SPECIES: *A. minima* Hofsten 1909, Scandinavia, Russia, acid waters; *A. agilis* Zacharias 1893, Europe, N. America, large waters and between sand grains.

TYPE SPECIES: *A. ecaudis* Perty 1850.

For further information see Beauchamp (1932), Carlin (1939, 1943).

ASPLANCHNIDAE

Lorica absent; foot present or absent (figs 86-91). Jaws incudate (fig. 4). Corona a simple girdle of arcs and tufts of cirri around head, and a very small area round mouth. Intestine and anus present or absent. Planktonic except for *Harringia*.

ASPLANCHNOPUS de Guerne 1888

One British and Irish species.

Foot ventral, short, 2 small toes (fig. 86); body sack-shaped; head with large apical area. No intestine and anus, gut ending at sack-shaped stomach; ovary horseshoe-shaped, contractile vesicle large. One cerebral and 2 lateral eyes. Length <1000 µm. Asexual eggs develop viviparously; resting eggs yellow with short bristles (fig. 86*c*); MALE <500 µm (fig. 134)—
(A. myrmeleo) **Asplanchnopus multiceps** (Schrank 1793)

Semi-planktonic in large and small waters, sometimes numerous, but very rare.

OTHER SPECIES (semi-planktonic): *A. dahlgreni* Myers 1934, N. America, acid water, shore region. *A. hyalinus* Harring 1913, N. America, brackish water.

TYPE SPECIES: *A. multiceps* (Schrank 1793) (described as *Brachionus multiceps).*

For further information on taxonomy see Harring (1913a), Myers (1934b), Beauchamp (1912), Wesenberg-lund (1930).

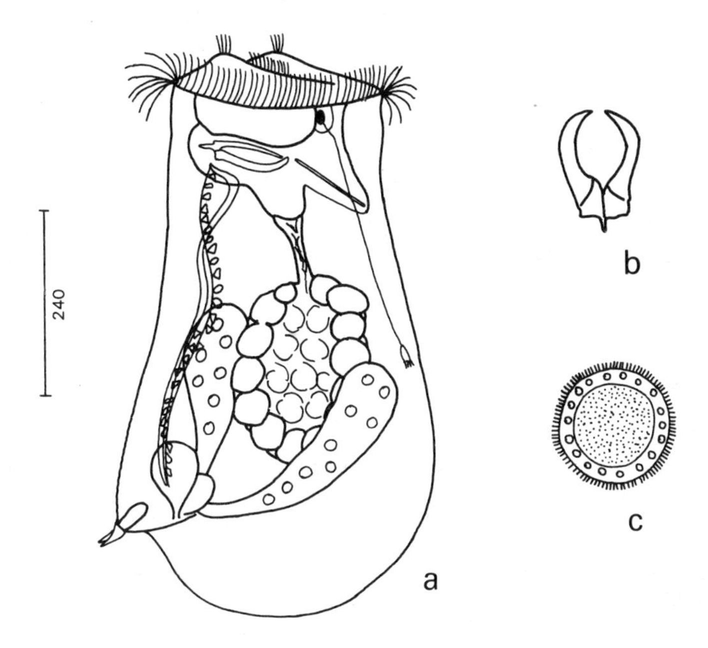

Fig. 86. *Asplanchnopus multiceps*. Female: *a*, lateral; *b*, jaws; *c*, sexual egg. (*a*, *b*, after Weber 1898, *c*, after Voigt 1957).

ASPLANCHNA Gosse 1850

No foot, intestine or anus; may be polymorphic; body sack-shaped, or with humps or projections, or bell-shaped (figs 87-90). One cerebral and 2 lateral eyes. Ovoviviparous.

1 Yolk gland round (figs 87, 88*a*↙)— ASPLANCHNA group **2**

— Yolk gland horseshoe-shaped (fig. 89*a*↙)— ASPLANCHNELLA group **3**

2 Pair of small foot glands present, opening by duct with uterus (fig. 87*a*, *b*↙). Large rotifers, <2000 µm. Body sack-shaped. Contractile vesicle very large and irregular in shape even when full, protonephridia compact with *c*. 50 flame bulbs each (fig. 87*a*). Rami serrated on inner edges (fig. 87*c*). Resting eggs with honeycomb shell; MALE July— **Asplanchna herricki** de Guerne 1888

Planktonic; sporadic in lakes and ponds; possibly not yet recorded from British Isles.

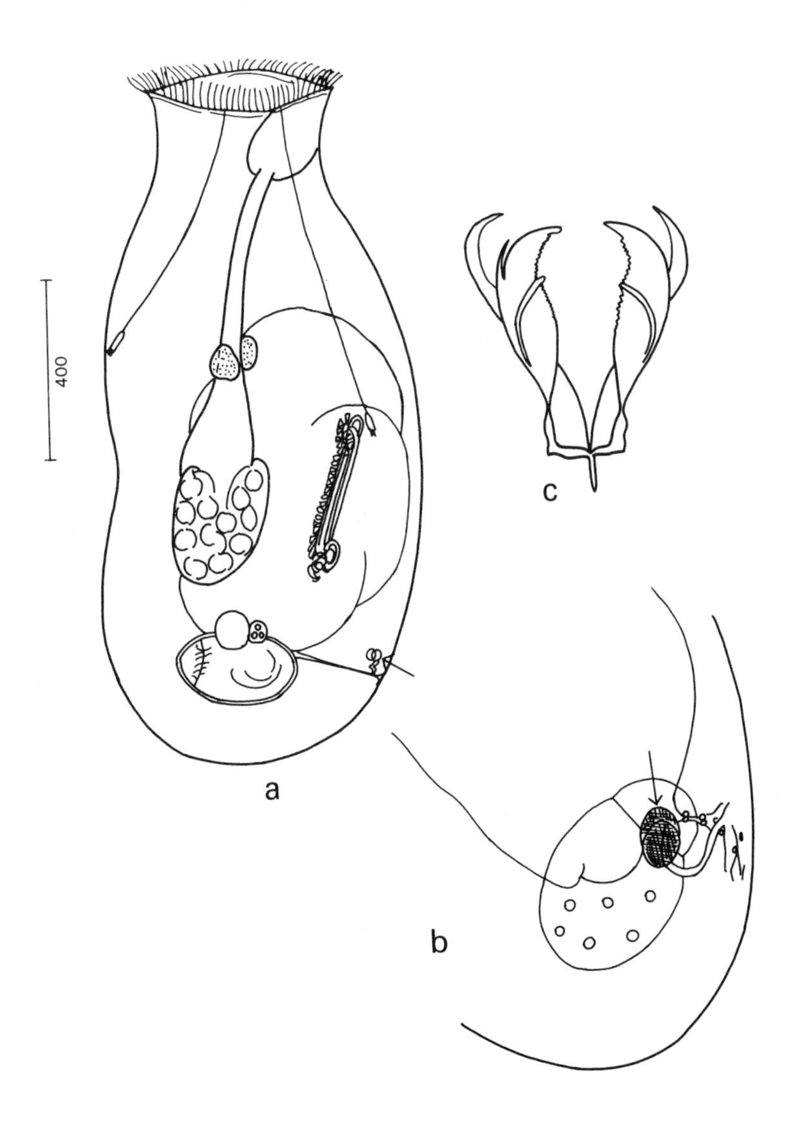

Fig. 87. *Asplanchna herricki.* Female: *a*, lateral; *b*, posterior end of body showing foot glands ↙ and duct; *c*, jaws. (*c*, after Wierzejski 1892).

— No foot glands. Length <1500 µm, usually *c*. 600 µm, sack-shaped (fig. 88). Contractile vesicle small, spherical when full, protonephridia compact with 4 flame bulbs each, uterovesicular duct opening ventrally (fig. 88*a*). Rami serrated on inner edges, broad at free ends (fig. 88*d*). Gastric glands rounded. Resting eggs with smooth laminated shell and orange oil drops, up to 150 µm (fig. 88*b*); MALE <500 µm, any time of year, (fig. 139*a*)—
Asplanchna priodonta Gosse 1850

Cosmopolitan, planktonic, perennial species, often numerous; size somewhat variable with temperature, larger in winter and in some localities; lakes and ponds and brackish water.

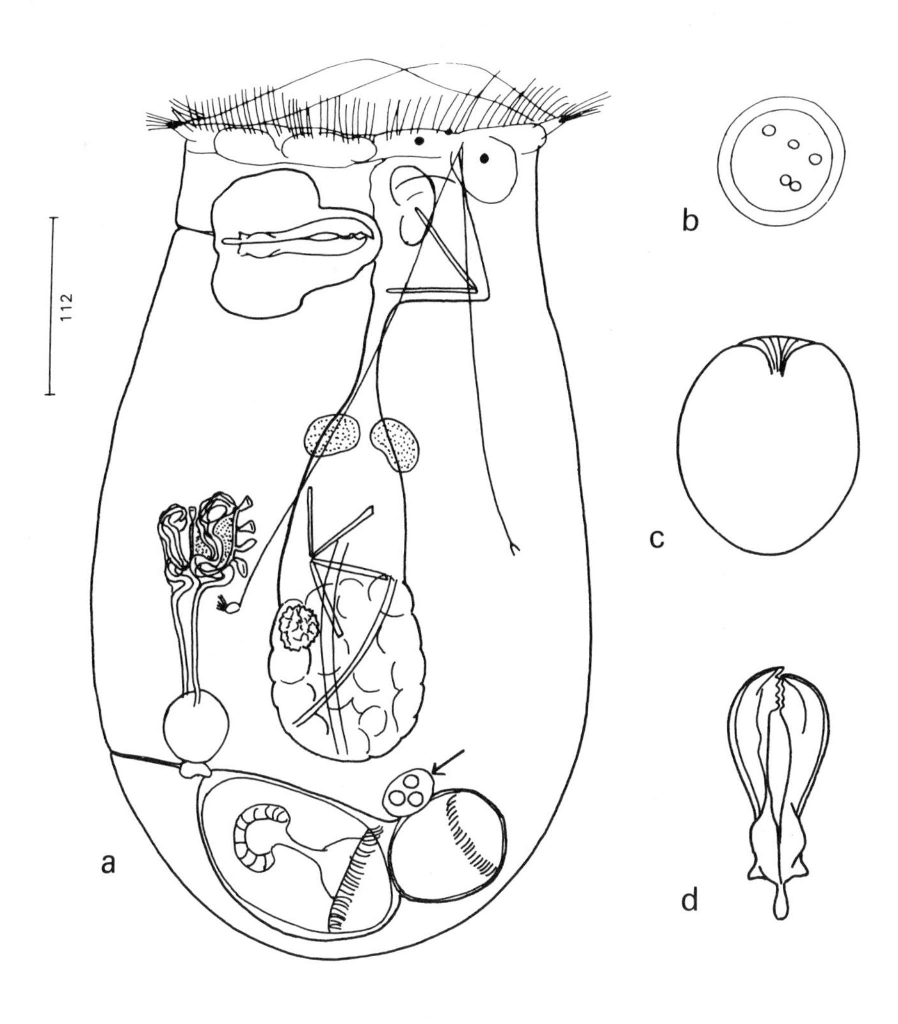

Fig. 88. *Asplanchna priodonta.* Female: *a,* lateral, ↙ yolk gland; *b,* sexual egg; *c,* with corona retracted; *d,* jaws.

3(1) Nuclei of yolk gland lobed (fig. 89*c*). Well-developed apophysis on ramus of jaws (fig. 89*d*, *e*↙). Male may have dorsal or lateral humps (fig. 139*c*). Female polymorphic, sack-shaped (fig. 89*a*), with humps (fig. 89*b*) or rarely bell-shaped. Length <2000 µm, usually *c.* 650 µm. Contractile vesicle irregular in shape, protonephridia long, extending up to head, each with 15-115 flame bulbs, uterovesicular duct opening posteriorly (fig. 89*a*). Gastric glands kidney-shaped. Resting eggs grey with many small lamellae or folds, up to 180 µm (fig. 89*f*); MALE <1200 µm, June-Aug., (fig. 139*b*, *c*)—
Asplanchna brightwelli sensu latissimo

Cosmopolitan, planktonic; freshwater generally and brackish water.

A number of species have been described which come under the umbrella name of *Asplanchna brightwelli* as used by Beauchamp (1951) and Gilbert (1968), and considerable variation between forms in characters used for identification has led to some confusion in the literature. However Gilbert (1973) states that *A. brightwelli* and *A. sieboldi* cannot be crossed so that there are at least 2 valid species in this group *A. intermedia* may be a third, or a variety of *A. brightwelli* (Gilbert 1968).

Inner margin of ramus with pronounced tooth (fig. 89*d*). Polymorphic, with sack-shaped, humped or bell-shaped forms. Flame bulbs from 20-40 each side for sack-shaped form to 80-115 for bell-shaped form— **A. sieboldi** (Leydig, 1854)
(A. amphora, A. ebbesborni, A. leydigi)

Inner margin of ramus with very small tooth. Little or no observable polymorphism; sack-shaped. Flame bulbs 15-20 each side—
A. brightwelli Gosse 1850

Inner margin of ramus without tooth (fig. 89*e*). Sack-shaped, no polymorphism. Flame bulbs *c.* 20 each side—
A. intermedia Hudson 1886

Fig. 89. *Asplanchna brightwelli* group. Female: *a*, sac-shaped form, lateral, ↙ yolk gland; *b*, humped form, ventral; *c*, yolk gland y, ovary o and egg e; *d*, *e*, jaws, *d*, tooth on inner margin of ramus well-developed; *e*, no tooth on inner margin of ramus; ↙ apophysis; *f*, sexual egg. (*b*, *e*, after Voigt 1957, *c*, after Waniczek 1930).

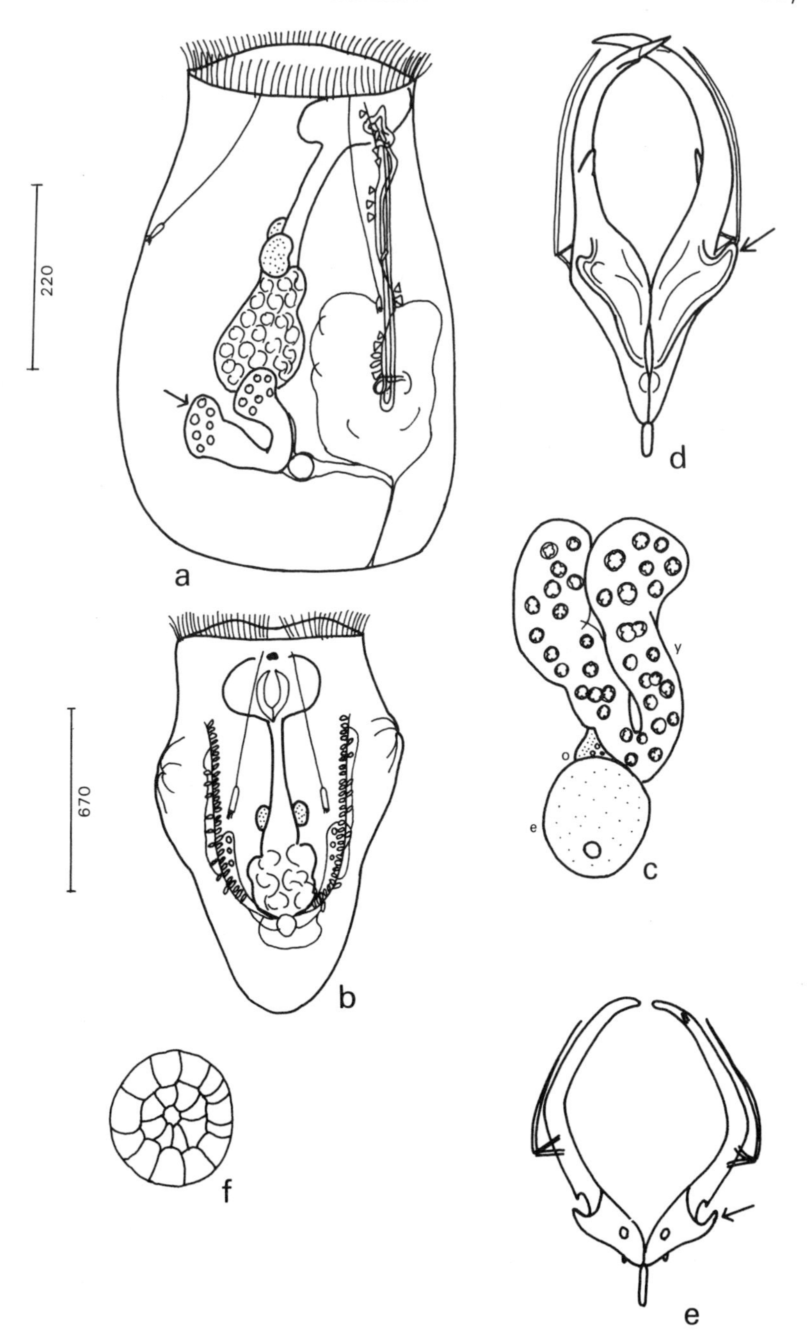
220
a
670
b
o
e
y
c
d
f
e

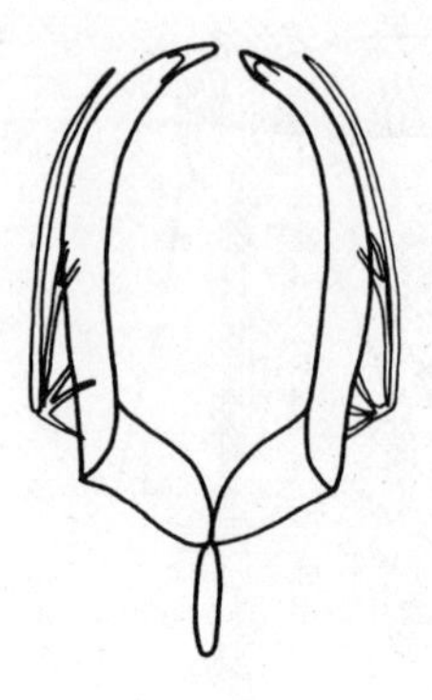

Fig. 90. *Asplanchna girodi.* Female: jaws. (After Voigt 1957).

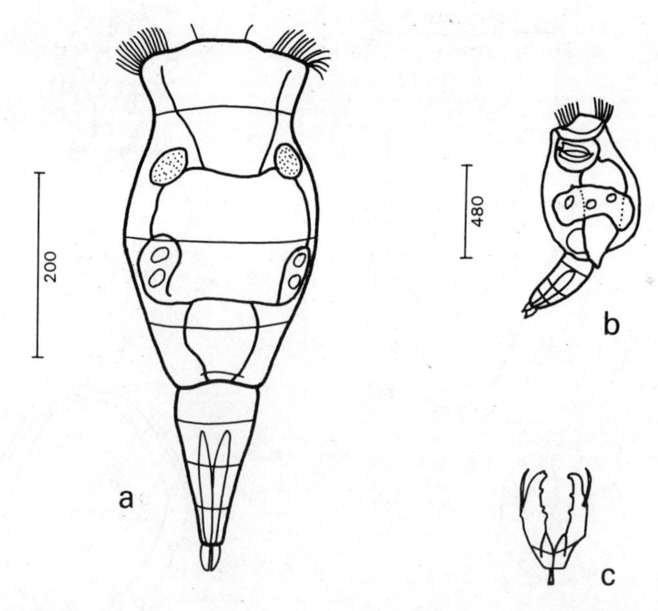

Fig. 91. *Harringia eupoda.* Female: *a,* ventral; *b,* lateral; *c,* jaws. (After Beauchamp 1912).

—(3) Nuclei of yolk gland rounded. No apophysis on ramus (fig. 90). Male never humped. Body sack-shaped. Length <700 µm. Protonephridia with 16 flame bulbs each. Resting eggs with honeycomb surface; MALE <400 µm—
Asplanchna girodi de Guerne 1888

Planktonic in ponds.

OTHER SPECIES: *A. silvestris* Daday 1902, America; inadequately described.

TYPE SPECIES: *A. priodonta* Gosse 1850.

For further information on the taxonomy and specific variability in the genus see Beauchamp (1951); for a description of some species see Waniczek (1930); for a description of the protonephridia in different species see Pontin (1964). Variations in form are described in Mitchell (1913), Gilbert & Thompson (1968), Gilbert (1973).

HARRINGIA de Beauchamp 1912

One British species

Foot stout, 2 small toes; body sack-shaped (fig. 91), in dorsal view similar to *Epiphanes senta* (fig. 41), but see lateral views (figs 91*b*, 41*b*). Ovary band-like. Corona small, 2 tufts of cilia. Length <700 µm— **Harringia eupoda** (Gosse, 1887)
(Asplanchna eupoda, Asplanchnopus eupoda, Dinops eupoda)

Probably mostly benthic in large rivers and temporary pools.

OTHER SPECIES: *H. rousseleti* de Beauchamp 1912, N. America, benthic.

TYPE SPECIES: *H. eupoda* (Gosse 1857) (described as *Asplanchna eupoda*).

For description of the species see Beauchamp (1912).

SYNCHAETIDAE

Lorica present or absent; foot present or absent. Jaws virgate, with or without hypopharynx muscle giving sucking action (fig. 5*a*, *b*). Corona reduced to small zone round mouth, and on anterior lobes or auricles, if present. Body sack-, bell- or cone-shaped. (Figs 92-105). A largely planktonic family.

POLYARTHRA Ehrb. 1834

No foot; body cylindrical or slightly flattened dorso-ventrally and roughly rectangular in dorsal and ventral views; body bears 4 groups of 3 feather-

or sword-shaped serrated blades or paddles, 2 groups dorso-lateral and 2 ventro-lateral, used in 'skipping' or 'rowing' movements, and arising from large lateral muscles. First generation animals from sexual eggs lack blades; see below (fig. 95*d*). In some species, a pair of small ventral 'finlets' is also present (figs 94*b*, 95*b*). Corona a single circle of cilia. Jaws virgate, large (fig. 5). Male very reduced (fig. 144).

Note on bladeless *Polyarthra* (have been described as separate genus *Anarthra*): resting eggs develop in most species as either f. *aptera* or f. *aptera reducta,* the eggs of which then develop into the typical form with blades.

F. *aptera:* body sack-shaped, very thin-walled; jaws present; 4 nuclei in ovary in *P. remata,* 8 in *P. dissimulans, P. dolichoptera* (fig. 95*d*).

F. *aptera reducta:* jaws absent, ventral side with deep fissure in position of mouth, stomach very reduced; coronal antennae absent; ovary very reduced, but well-developed egg already present in body cavity. In *P. euryptera, P. vulgaris.*

NOTES ON USE OF KEY TO SPECIES OF POLYARTHRA:

1. Breadth of blades should be measured, by micrometer, ensuring that the blade is flat, i.e. presenting its broad side, not its edge, to view.
2. Presence or absence of ventral finlets (figs 94*b*, 95*b*) can only be determined for certain in specimens presenting lateral view.

If a number of specimens can be placed under a coverslip together, by chance some usually present lateral view and some have blades lying flat and clearly showing. Preserved specimens in a deep watch-glass i.e. uncompressed, show ventral finlets standing out from body.

1 Blades feather-like, up to 60 µm broad (figs 92-94)— **2**

— Blades serrated sword-shaped, up to about 15 µm broad (figs 95-97)— **4**

2 Blades broad, 30-60 µm (fig. 92). Nuclei in yolk gland 12. Length of blades less than length of body. No ventral finlets. Body <250 µm. Resting eggs autumn, usually double-shelled, with folded skin between shells (fig. 92*b*); first generation f. *aptera reducta*—
Polyarthra euryptera Wierzejski 1893

Planktonic; lakes and ponds in summer, warm stenotherm, eutrophic waters.

— Blades less than 30 µm broad. Nuclei in yolk gland 8— **3**

Fig. 92. *Polyarthra euryptera.* Female: *a*, dorsal; *b*, sexual egg.

3 Blades *c*. 20-30 µm broad (fig. 93). No ventral finlets. Length of blades less than body length. Body <180 µm. Resting egg with either single shell with thickening, or double shell; outer shell smooth, transparent, up to 70×100 µm (fig. 93*b*); 1st generation unknown—
P. major Burckhardt 1900
(*P. trigla* var. *major*, *P. platyptera* var. *major*).

Planktonic; frequent in oligotrophic waters June-Nov., usually with other species; eurytherm.

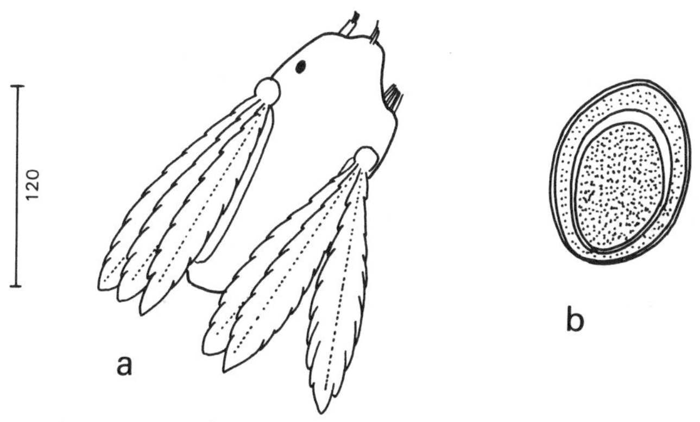

Fig. 93. *Polyarthra major.* Female: *a*, lateral; *b*, sexual egg. (*b*, after Nipkow 1952).

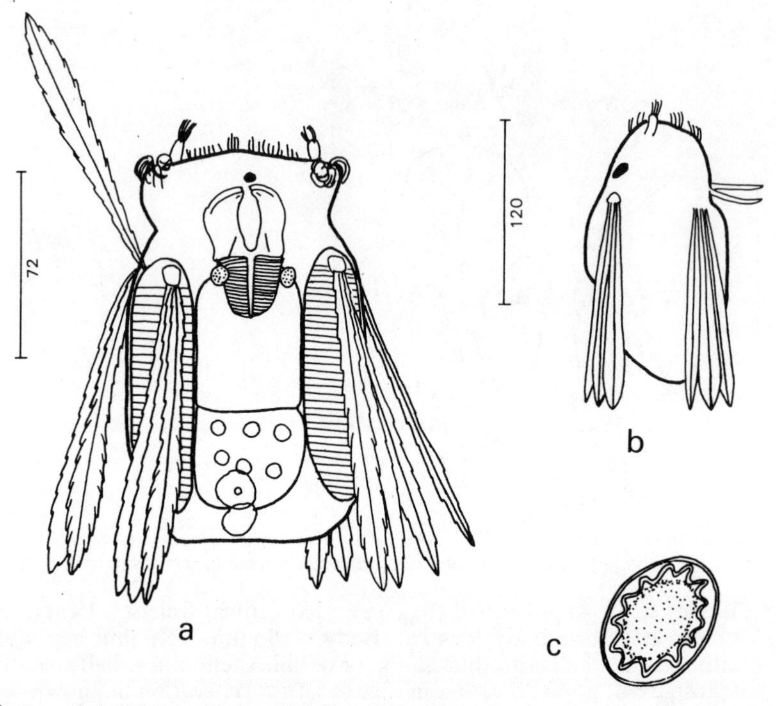

Fig. 94. *Polyarthra vulgaris*. Female: *a*, ventral; *b*, lateral, showing ventral finlets; *c*, sexual egg. (*c*, after Nipkow 1952).

— Blades *c*. 8-20 µm broad, but variable, grading from feather-shape towards narrower sword-shape (fig. 94) (see *P. dolichoptera* fig. 95). Ventral finlets present (fig. 94*b*). Body <165 µm. Blades equal to body length or shorter. Resting egg double-shelled with folded skin between shells (fig. 94*c*); 1st generation f. *aptera reducta* (male fig. 144)— **Polyarthra vulgaris** Carlin 1943 *(P. trigla)*

Cosmopolitan, planktonic; larger waters, canals; numerous in summer, sporadic in winter; eurytherm, possibly in oxygen-rich waters.

A variable species grading towards *P. dolichoptera*.

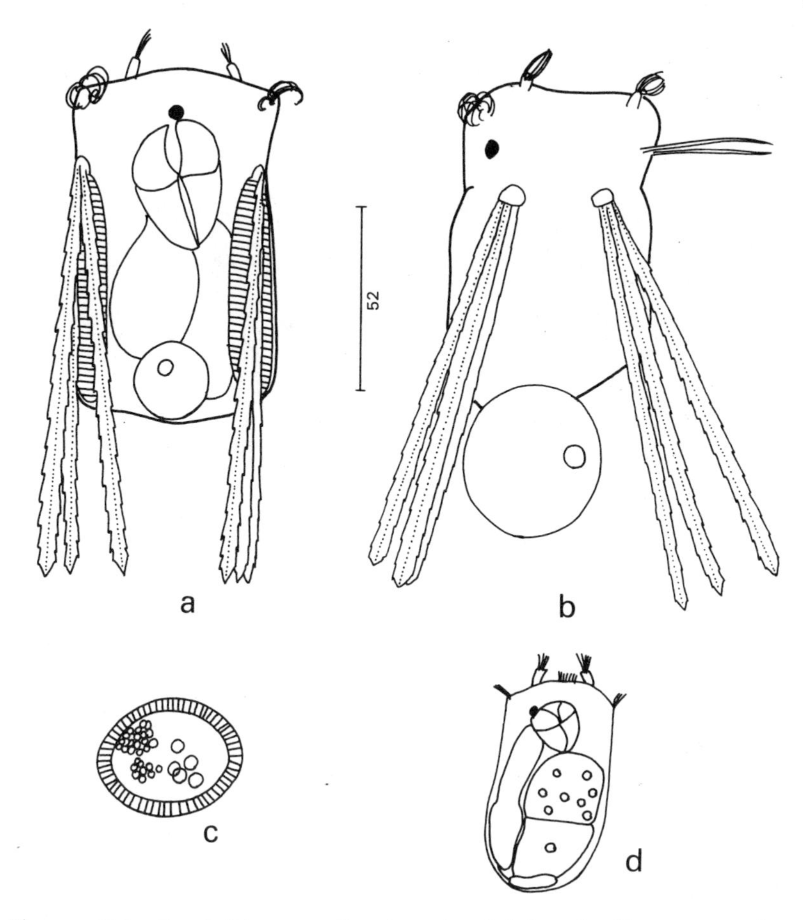

Fig. 95. *Polyarthra dolichoptera.* Female: *a*, dorsal; *b*, lateral, with egg; *c*, sexual egg; *d*, forma *aptera*.

4(1) Ventral finlets present (fig. 95*b*). Nuclei in yolk gland 8. Blades long, longer than body, easily extending past body end, <220 μm long and <14 μm broad. Body <145 μm. Resting eggs with outer shell with spines, up to 72×56 μm (fig. 95*c*); 1st generation f. *aptera* (fig. 95*d*)— **P. dolichoptera** Idelson 1925
(*P. platyptera* var. *dolichoptera*)

Planktonic; large and small lakes and brackish waters; possibly in colder seasons.

— Ventral finlets absent. Nuclei in yolk gland 4— **5**

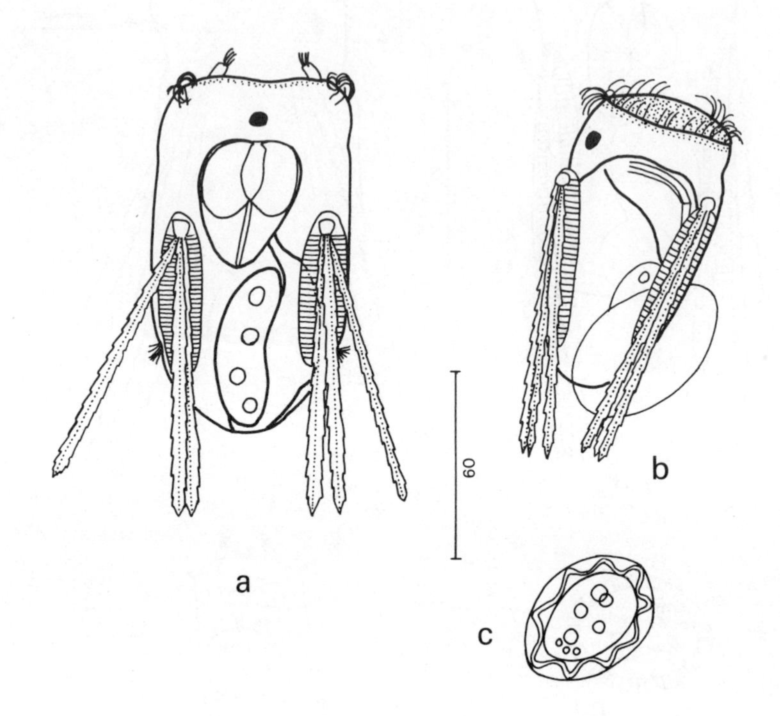

Fig. 96. *Polyarthra remata.* Female: *a,* dorsal; *b,* lateral, with egg; *c,* sexual egg.

5 All blades the same length, rather longer than body and 7-8 µm broad (fig. 96). Body <120 µm. Coronal area often coloured reddish or purplish-red. Resting egg dark-grey then orange-red, double-shelled with folded skin between shells, up to 60×44 µm, Oct. (fig. 96*c*); 1st generation f. *aptera*— **Polyarthra remata** Skorikov 1896

Cosmopolitan, planktonic; small waters, lakes, canals, alkaline-neutral waters; June-Oct., sometimes Jan., maximum Sept.

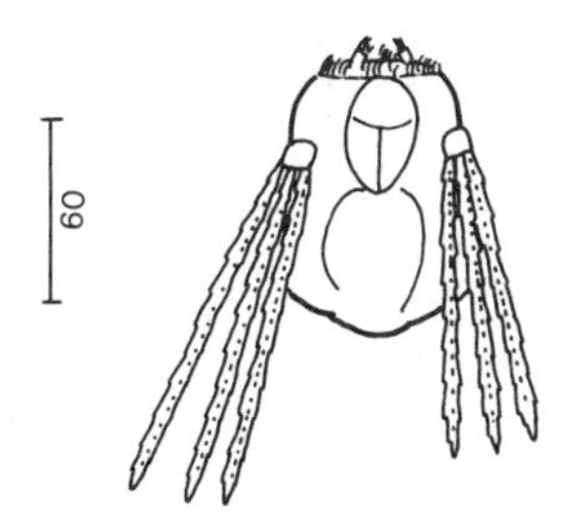

Fig. 97. *Polyarthra minor*. Female. (After Voigt 1957).

— Left dorsal blades longer than rest (fig. 97). Blades up to 4 µm broad. Body <90 µm. Males and resting eggs unknown—
(P. platyptera var. *minor)* **P. minor** Voigt 1904

Ponds and acid waters; between *Sphagnum* and in open water.

OTHER SPECIES (of doubtful status): *P. longiremis* Carlin 1943, similar to *P. vulgaris* and *P. dolichoptera,* has blades about body length and up to 8 µm broad, body <220 µm; small ponds, May-June, recorded from Shropshire. *P. dissimulans* Nipkow 1952, also in the *vulgaris-dolichoptera* group, has blades reaching to about end of body, 8-12 µm broad, body <176 µm. Resting egg double-shelled, reddish, no spines, 1st generation f. *aptera;* Switzerland, not recorded from Britain and Ireland.

TYPE SPECIES: *P. vulgaris* Carlin 1943

For taxonomic literature see Albertova (1960), Amrèn (1964), Carlin (1943), Nipkow (1952), Pejler (1957), Ruttner-Kolisko (1959). A key to the genus is given in Bartoš (1951a).

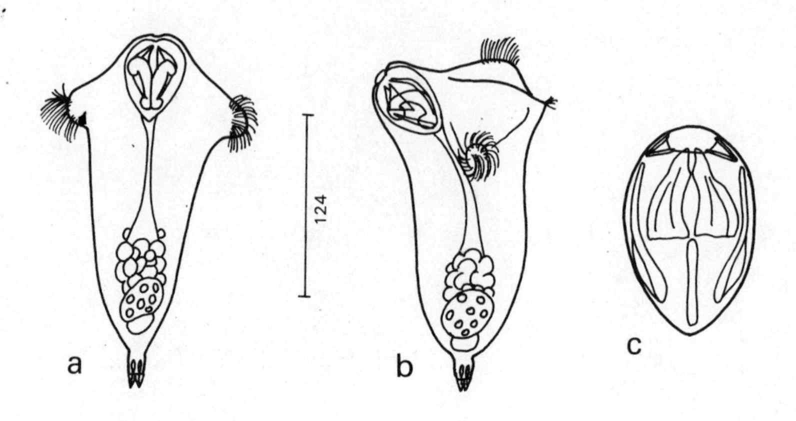

Fig. 98. *Synchaeta calva.* Female: *a,* ventral; *b,* lateral; *c,* jaws. (After Ruttner-Kolisko 1970).

SYNCHAETA Ehrb. 1832

Foot short or fairly short, toes short, usually 2; body bell- or cone-shaped, cuticle very thin and transparent; head more or less convex, usually bearing 4 characteristic stiff sensory bristles or styli arising from triangular prominences. (Figs 98-103). Corona of groups of cirri, also cirri on laterally-extending or dependent 'ears' or auricles. Cerebral eye usually present, single or double, red or purple, and often pigment spots on head. Virgate jaws large and conspicuous, with hypopharynx (fig. 5*a*, *b*). Very active swimmers.

1 With 4 styli on head, with or without bristle-bearing tentacles (figs 99-103)— **2**

— No styli or tentacles on head (fig. 98). Body conical when swimming, vase-shaped under coverslip; foot parallel-sided, set off from body, 2 small toes. Length <250 µm. No eye. Jaws large, protruding ventrally in a conical bulge, unci with double pointed tips (fig. 98c). Lateral antennae very small, $\frac{2}{3}$ of the way down body—
Synchaeta calva Ruttner-Kolisko 1969
(?True species or aberrant form)

Planktonic in lakes.

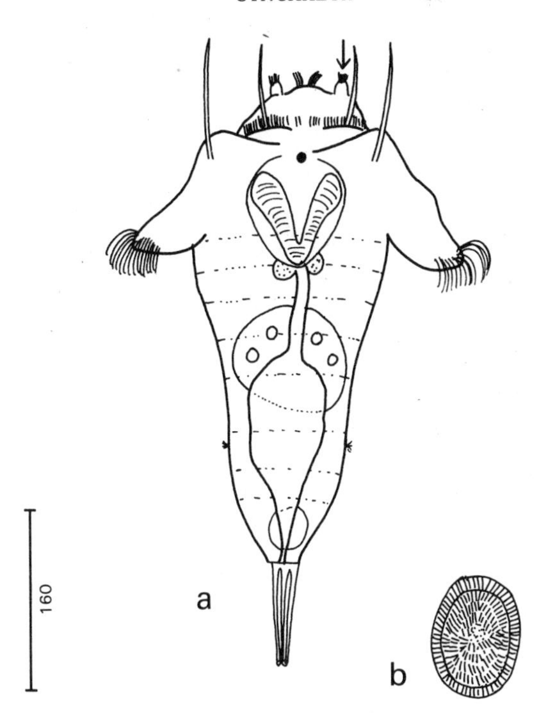

Fig. 99. *Synchaeta grandis*. Female: *a*, dorsal, ↙ tentacle, *b*, sexual egg. (*a*, after Rousselet 1902, *b*, after Voigt 1957).

2 Foot relatively long, more than twice as long as broad, clearly set off from body (figs 99, 100)— **3**

— Foot shorter than this (figs 101-103)— **4**

3 Body a slender cone shape, constricted in the middle (fig. 99); head convex, with or without 2 small bristle-bearing tentacles (fig. 99*a*↙). Length <600 µm. Unci of jaws with pointed tips. Eye single. Asexual eggs yellow; resting eggs with slightly curved spines or with bumps, grey, 100×85 µm, July, (fig. 99*b*); males unknown—
S. grandis Zacharias 1893

Planktonic; lakes and moor pools, oligotrophic and dystrophic waters; rare.

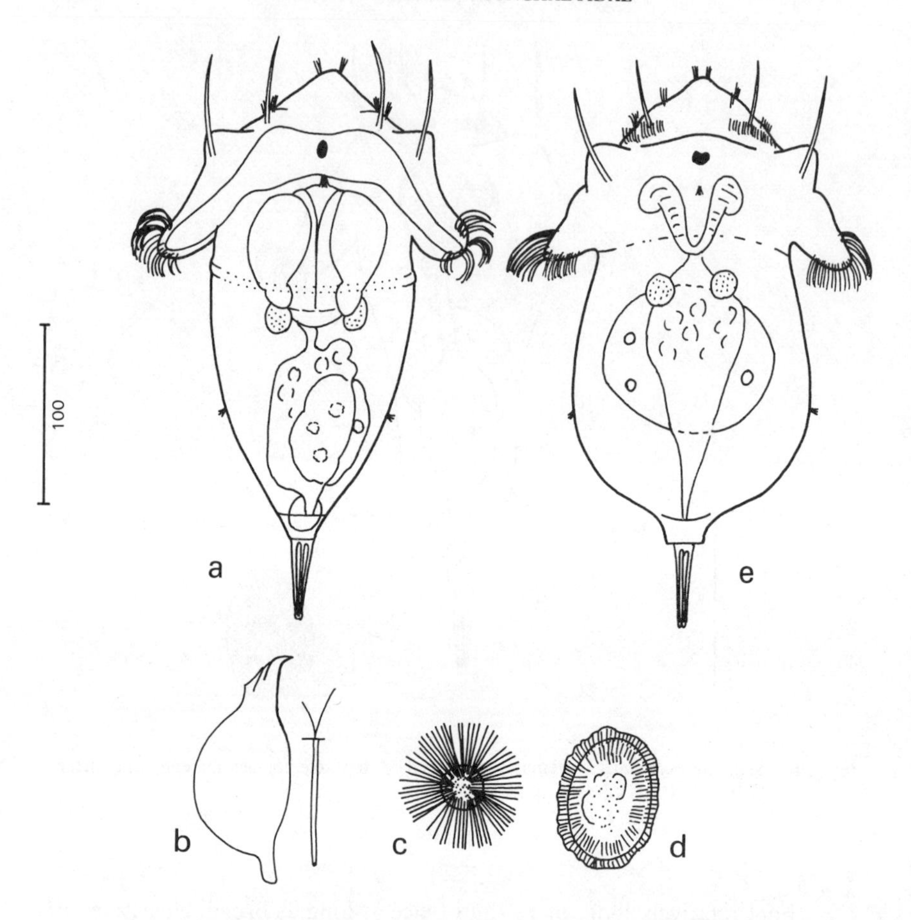

Fig. 100. *Synchaeta*. Female: *a-d, S. stylata; a,* dorsal; *b,* uncus and manubrium; *c,* asexual egg; *d,* sexual egg; *e, S. longipes,* dorsal. (*d,* after Nipkow 1961, *e,* after Rousselet 1902).

— Body bell-shaped, or a plump to slender cone (fig. 100); head convex, no apical tentacles. Length <310 μm. Unci of jaws with pointed tips (fig. 100*b*). Eye single. Asexual eggs with oil drops and long thin bristles; resting eggs with short rods in shell (fig. 100*c*, *d*); males unknown— **Synchaeta longipes** Gosse 1887 and **S. stylata** Wierzejski 1893

Planktonic, cosmopolitan, but sporadic in lakes and ponds in summer and autumn (rarely in brackish water).

These 2 species are very similar and seem to be separated only on the shape of the body (*S. longipes* plumper and broader at the level of the lateral antennae, *S. stylata* more slender and broader towards the anterior end) and the foot (*S. longipes* straight-sided, cylindrical, narrow, *S. stylata* tapering) (fig. 100*a*, *e*).

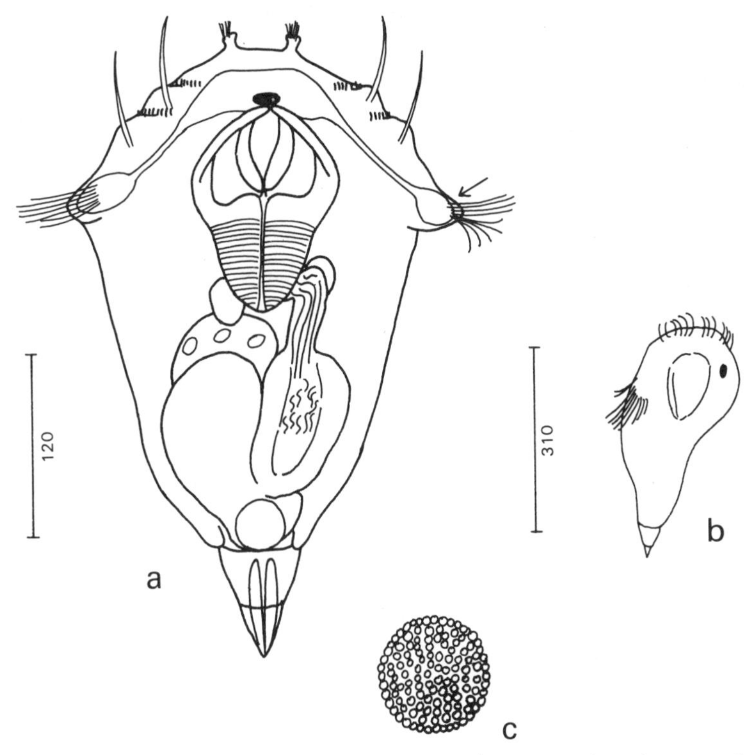

Fig. 101. *Synchaeta pectinata.* Female; *a*, dorsal, ↙ auricle; *b*, lateral; *c*, sexual egg. (*c*, after Rousselet 1902).

4(2) Two bristle-bearing apical tentacles (fig. 101*a*). Body bell-shaped, head convex. Length <510 µm. Eye single. Unci of jaws with pointed tips (fig. 101*a*). Asexual eggs with oil drops and gelatinous case; resting eggs with blisters or short spines (fig. 101*c*); MALE <160 µm (fig. 135*a*)— **S. pectinata** Ehrb. 1832

Planktonic, perennial species, numerous in winter and spring; also warm water form with maxima in summer and autumn (maxima possibly in falling temperatures). Cosmopolitan, large and small waters, canals; also brackish waters.

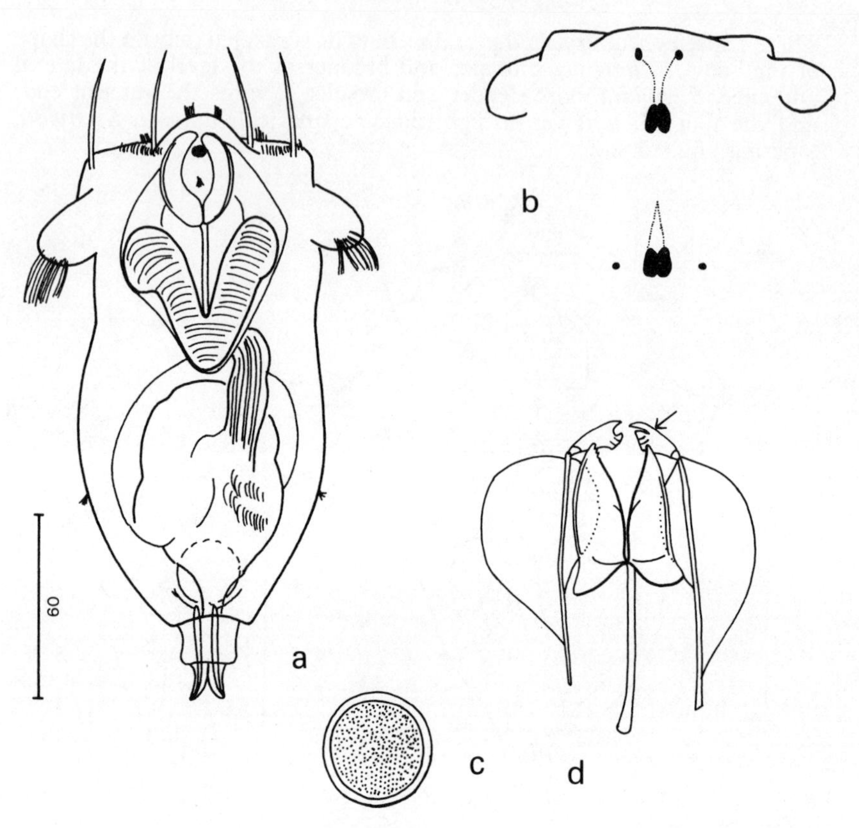

Fig. 102. *Synchaeta oblonga*. Female: *a*, dorsal; *b*, dorsal view of head showing variable arrangement of eye pigment; *c*, sexual egg; *d*, jaws, ↙ uncus with comb-like edge.

— No apical tentacles (figs 102, 103). Unci with comb-like ends (fig. 102*d*↙). Cerebral eye double (fig. 102*b*)— **5**

5 Body bell-shaped, head convex (fig. 102). Length <250 μm. Lateral antennae $\frac{2}{3}$ of the way down body (fig. 102). Eggs carried for a short time; resting eggs with or without short rods in shell. MALE <100 μm, spring (fig. 135*c*)—

Synchaeta oblonga Ehrb. 1832.

Planktonic; lakes and ponds, moor waters and canals, frequent in winter and spring, also in higher temperatures.

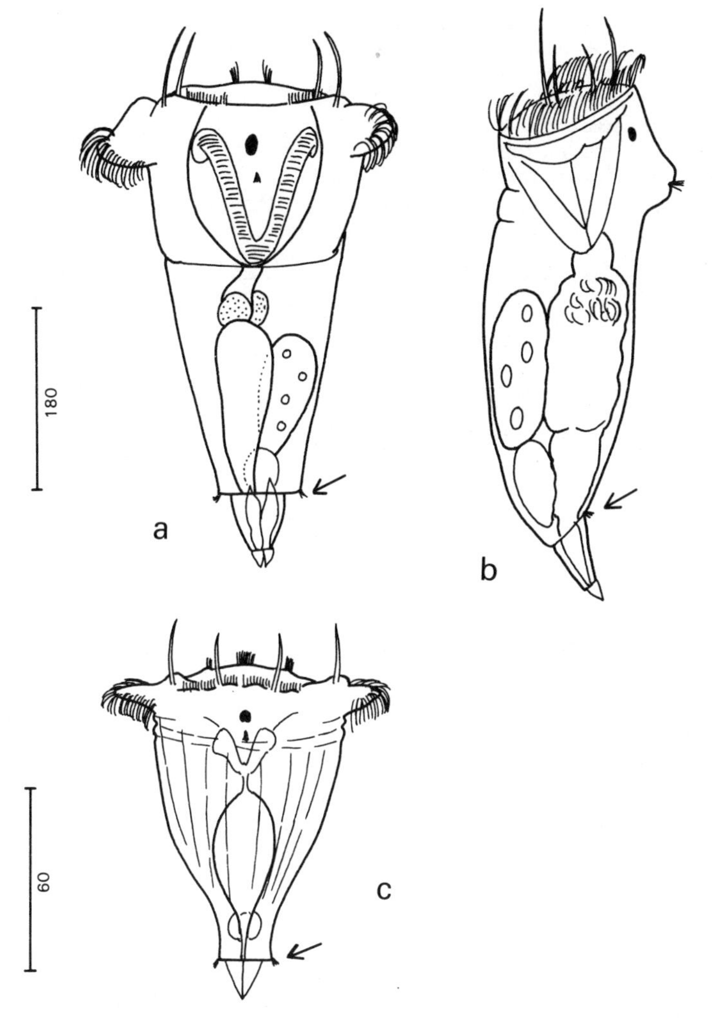

Fig. 103. *Synchaeta*. Female: *a, b, S. tremula, a,* dorsal; *b,* lateral; *c, S. kitina,* dorsal; ↙ lateral antennae. (*b,* after Hollowday personal communication, *c,* after Rousselet 1902).

— Body cone-shaped, head more or less flat (fig. 103). Lateral antennae at base of foot (fig. 103↙). Eggs sometimes carried or laid on diatoms; resting eggs with small spines. MALE <120 μm, May and Oct. (fig. 135*b*)— **S. tremula** (Müller 1786) and **S. kitina** Rousselet 1902

Widespread in large and small waters (also inland brackish waters); littoral, between plants and sand, and in plankton; frequent in winter and spring.

These 2 species were considered to be identical by Wesenberg-Lund (1930), and are united by Ruttner-Kolisko (1974). Pejler (1957) considered *S. kitina* to be the extreme littoral form, but Nipkow (1961) found it as a summer plankton form and Galliford (personal communication) also considers it as more truly planktonic than *S. tremula*. (Fig. 103). *S. kitina* is very small, *c.* 100 µm, maximum 136 µm; *S. tremula c.* 150-300 µm.

OTHER SPECIES: freshwater non-British planktonic species: *S. lakowitziana* Lucks 1930, cold stenotherm; *S. verrucosa* Nipkow 1961, L. Zurich; *S. pachypoda* Jaschnow 1922, L. Baikal; *S. wesenberg-lundi* Pejler 1957. Marine and brackish species: Ruttner-Kolisko (1974) lists 21, of which 8 or 9 are British and Irish.

TYPE SPECIES: *S. pectinata* Ehrb. 1832.

Other literature on the taxonomy of the genus and descriptions of species includes Amrèn (1964), Bērziņš (1960), Carlin (1943), Donner (1959), Nipkow (1961), Parise (1961), Pejler (1957), Pourriot (1965), Rousselet (1902), Ruttner-Kolisko (1970), Wesenberg-Lund (1930).

PLOESOMA Herrick 1885

Foot partly wrinkled, 2 toes; lorica sack- or cone-shaped, with 'bubbly' surface or with ridges (figs 104, 105). Head with 2 finger-like palps. Eye present.

1 Lorica ridged and with ventral furrow (fig. 104*a-c*); head-shield present, without teeth. Foot of medium length. Eye red. Length <300 µm. Egg 270×225 µm; males unknown— **Ploesoma truncatum** (Levander 1894)

Lakes and ponds; widespread, summer form, in water at higher temperatures; semi-planktonic, also between plants.

Similar, but with 3 spines on head-shield (fig. 104*e*)— *(P. lynceus)* **P. triacanthum** (Jennings 1894)

or with 1 spine on head-shield (fig. 104*d*)— **P. lenticulare** (Herrick 1885)

— Lorica with blisters or vacuoles ventrally, giving a 'bubbly' appearance (fig. 105); no ventral furrow in lorica; no head-shield. Foot long. Single eye red or black. Length <612 µm. Eggs with transparent shell, up to 326×300 µm; resting eggs grey, with thick yellow wrinkled shell, 230×190 µm (fig. 105*c*, *d*); MALE <160 µm (fig. 138)— **P. hudsoni** (Imhof 1891)

Often numerous in lakes which are warm in summer, rare in ponds; planktonic.

TYPE SPECIES: *P. lenticulare* Herrick 1885.

For further information on the genus see Nipkow (1961), Pourriot (1965), Wesenberg-Lund (1930).

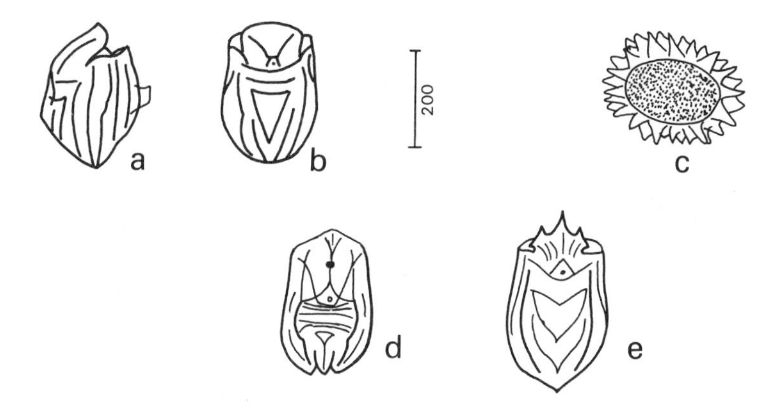

Fig. 104. *Ploesoma*. Female: *a-c*, *P. truncatum*, *a*, lorica, lateral; *b*, lorica, dorsal; *c*, egg; *d*, *P. lenticulare*, lorica, dorsal; *e*, *P. triacanthum*, lorica, dorsal. (*a*, *b*, after Levander 1894, *c*, after Nipkow 1961, *d*, *e*, after Bartoš 1959).

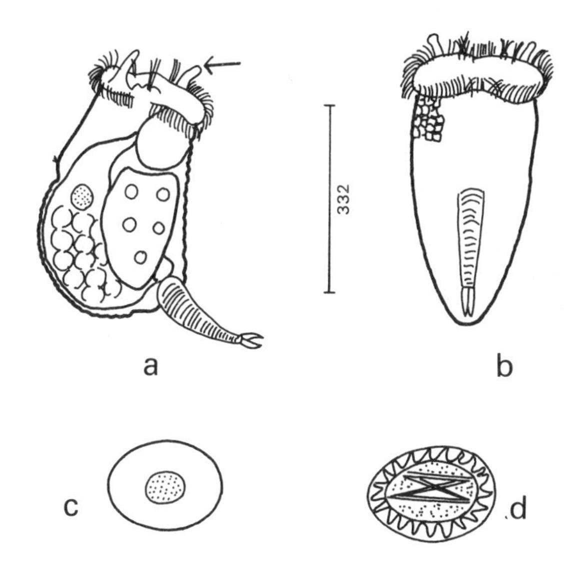

Fig. 105. *Ploesoma hudsoni*. Female: *a*, extended, lateral view, ↙ palp; *b*, ventral; *c*, asexual egg; *d*, sexual egg. (*a*, *c*, after Voigt 1957, *b*, after Edmondson 1959, *d*, after Nipkow 1961).

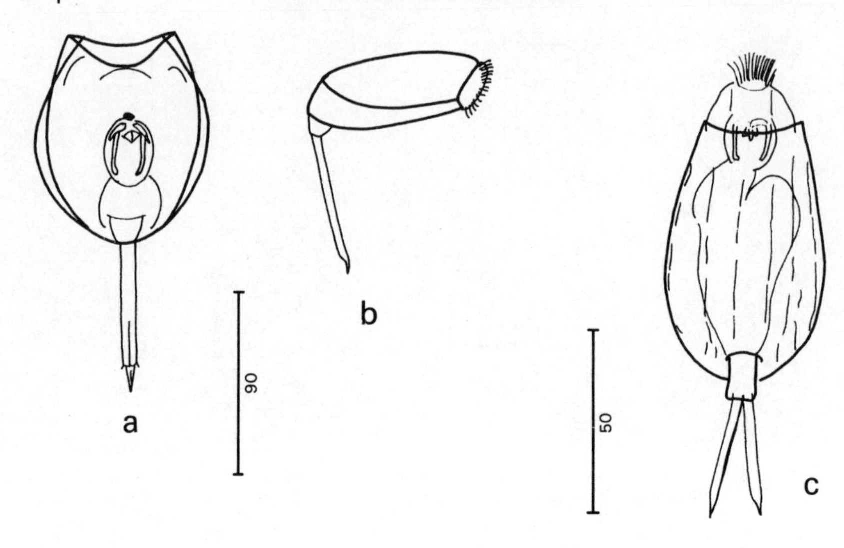

Fig. 106. *Lecane*. Female: *a, b, L. lunaris, a,* ventral; *b,* lateral; *c, L. subtilis,* ventral.

LECANIDAE

LECANE Nitzsch 1827

Foot very short, 1 or 2 sections, toes long, 2 or fused to 1. Lorica flattened dorsoventrally, oval, pear- or shield-shaped, with dorsal and ventral plates; head opening broad and shallow, with lateral edges prolonged into angles or short spines; posterior end rounded or extended into a process. Single eye. Jaws malleate. (Fig. 106). A littoral genus, occurring between plants.

TYPE SPECIES: *L. luna* (Müller 1776) (described as *Cercaria luna).*

For information on the genus see Carlin (1939), Harring & Myers (1926), Hauer (1929), Voigt (1957).

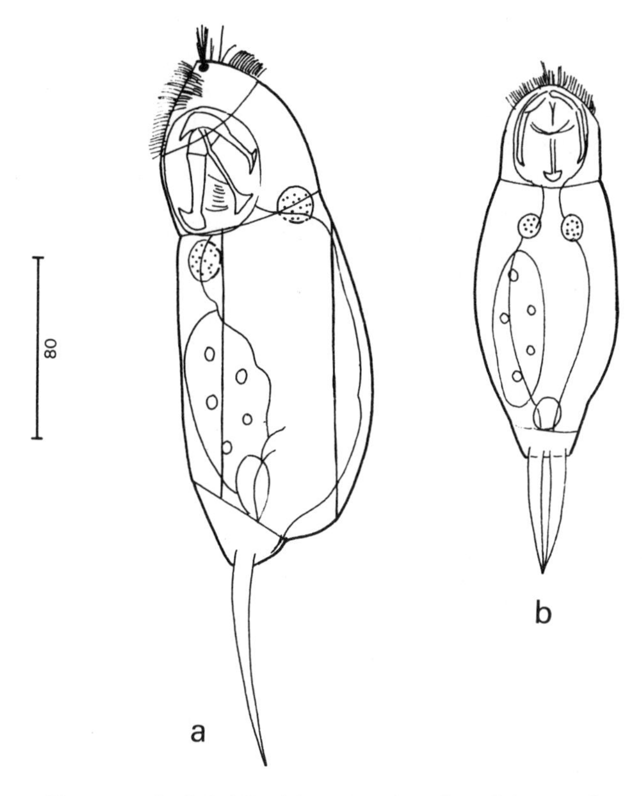

Fig. 107. *Cephalodella gibba*. Female: *a*, lateral; *b*, ventral.

NOTOMMATIDAE

CEPHALODELLA Bory de St Vincent 1826

Foot very short, toes short to long, usually curved. Lorica delicate, of several pieces, and with lateral cleft. Body slightly curved towards ventral side, and somewhat compressed laterally. Jaws virgate. (Fig. 107). A littoral genus, occurring amongst vegetation.

TYPE SPECIES: *C. catellina* (Müller 1786) (described as *Cercaria catellina)*
For further information see Donner (1950), Harring & Myers (1924), Wulfert (1938).

TESTUDINELLIDAE

Lorica present or absent; foot present or absent; toes absent (figs 108-116). Jaws malleoramate (fig. 7*b*). Corona a double circumapical ring with mouth between rings. Planktonic and periphytic species.

TESTUDINELLA Bory de St Vincent 1826

A number of British and Irish species which are almost entirely periphytic; one species which is also found in plankton.

Lorica strongly flattened dorso-ventrally, almost circular in outline (fig. 108); foot opening circular, almost in middle of ventral side; foot wrinkled, toes replaced by circle of cilia. Length <135 μm (male fig. 129)— **Testudinella patina** (Hermann 1783)

Cosmopolitan; fresh and brackish waters.

OTHER SPECIES: British, periphytic: *T. mucronata* (Gosse 1886), *T. truncata* (Gosse 1886), *T. parva* & *T. incisa* (Ternetz 1892), *T. reflexa* (Gosse 1887). Periphytic and epizoic on *Asellus: T. clypeata* (Müller 1786), *T. caeca* Parsons 1892), *T. elliptica* (Ehrb. 1834).

TYPE SPECIES: *T. clypeata* (Müller 1786) (described as *Brachionus clypeatus).*

For further information on the genus see Bartoš (1951b), Gillard (1947), Myers (1934a), Ruttner-Kolisko (1974) and Seehaus (1930).

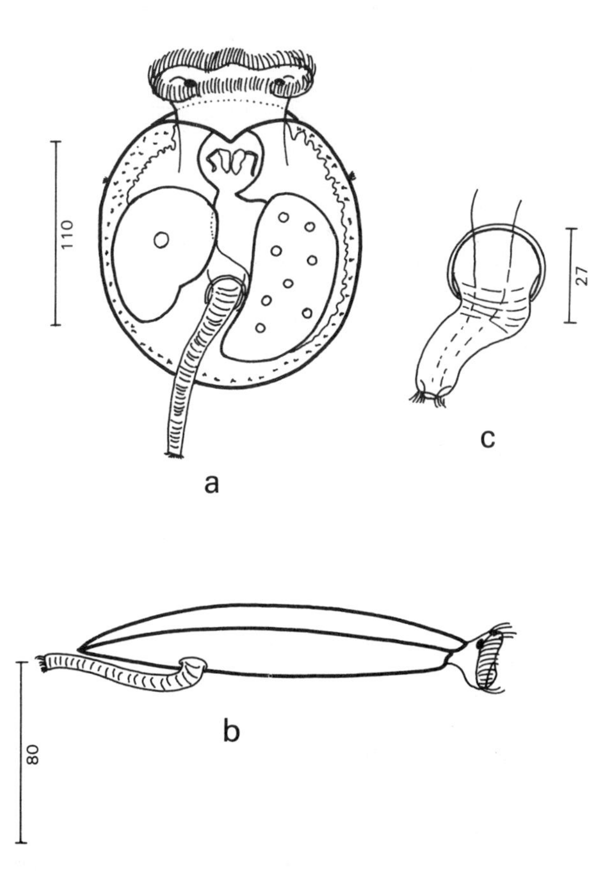

Fig. 108. *Testudinella patina.* Female: *a,* extended, ventral; *b,* lateral; *c,* foot opening, foot, and circle of cilia in place of toes.

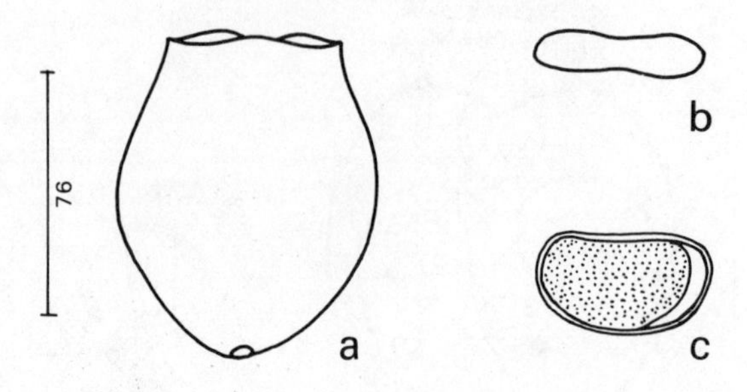

Fig. 109. *Pompholyx complanata.* Female: *a*, lorica, dorsal; *b*, cross-section; *c*, sexual egg. (*a*, *b*, after Voigt 1957, *c*, after Rylow 1935).

POMPHOLYX Gosse 1851

No foot; lorica oval or shield-shaped, flattened dorso-ventrally or not, no spines (figs 109, 110). Eggs carried on retractile threads secreted by a gland, the threads passing out through a conspicuous cloacal opening at posterior end of lorica (fig. 110*e*). Two eyes.

1 Lorica flattened dorso-ventrally, cross-section as fig. 109*b*. Length <90 μm. Resting eggs dark, smooth, up to 70×40 μm (fig. 109*c*); males unknown— **Pompholyx complanata** Gosse 1851

Widespread, but more rare than *P. sulcata*; planktonic in eutrophic ponds and lakes.

— Lorica not flattened dorso-ventrally, grooved to give cross-section as fig. 110*b*. Length <126 μm. Eggs round, up to 75 μm, carried for some time on retractile thread from secretory gland (fig. 110*e*, *f*); resting eggs dark, smooth, double-shelled, up to 80×70 μm (fig. 110*d*); MALE <70 μm (fig. 142)— **P. sulcata** Hudson 1855

Planktonic; widespread, usually sporadic but sometimes numerous in eutrophic ponds and lakes; also in brackish water.

TYPE SPECIES: *P. complanata* Gosse 1851.

For literature on the genus see Bartoš (1951b), Leissling (1924), Remane (1933); for egg-carrying see Sudzuki (1955a).

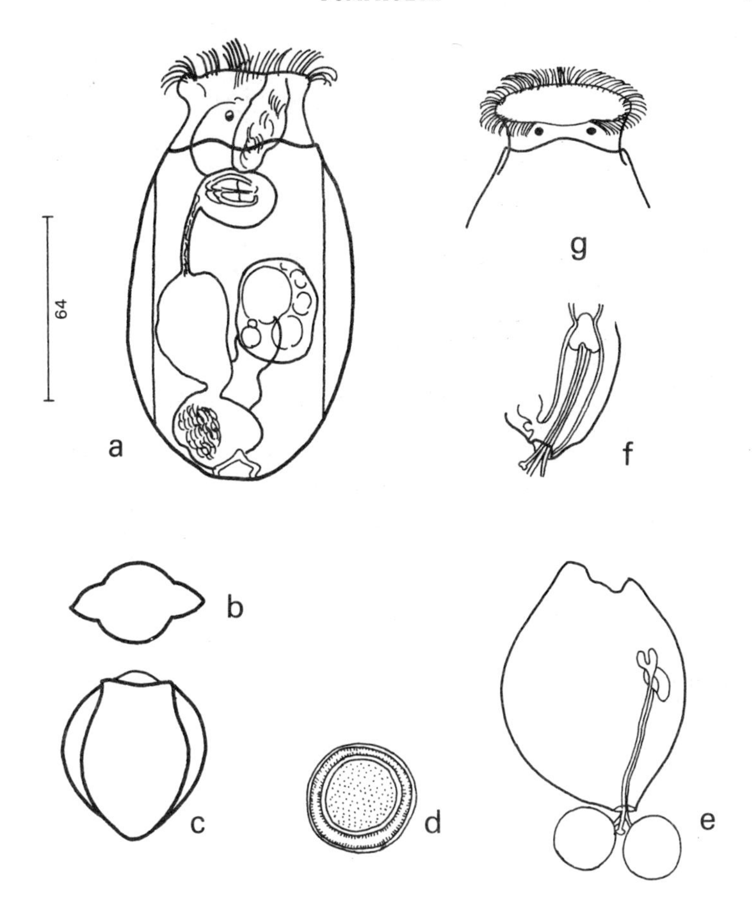

Fig. 110. *Pompholyx sulcata.* Female: *a*, ventral; *b*, lorica cross-section; *c*, lorica, ventral; *d*, sexual egg; *e*, lateral view showing eggs carried on a thread; *f*, posterior end showing secretory gland; *g*, anterior end, dorsal. (*d*, *e*, *f*, after Sudzuki 1955a, *g*, after Leissling 1924).

HEXARTHRA Schmarda 1854

One British and Irish species.

No foot; body cone- or bell-shaped, colourless or yellowish, reddish, sometimes brick-red or blue. Six bristle-bearing arm-like processes: 1 dorsal, 1 ventral, 2 latero-dorsal and 2 latero-ventral, each ending in a fan of feathered bristles, ventral arm especially long (fig. 111); 2 club-shaped caudal processes at posterior end (fig. 111*b*↙). Underlip present (fig. 111). Length <400 µm. Eggs, including male eggs, carried; resting eggs brown, shell covered with short, thick, yellowish tubes which look like prisms under mutual pressure, and with a tube-free area forming a ring round the egg equator (fig. 111*c*); MALE <50 µm, July and Aug. (fig. 130)—

Hexarthra mira (Hudson, 1871)

Cosmopolitan, planktonic; lakes and ponds which are warm in summer, also moor waters and brackish water.

OTHER SPECIES (all planktonic): *H. intermedia* Wisniewski 1929 (syn. *H. insulana)*, Europe, tropics; *H. propinqua* (Bartoš 1948), *H. mollis* (Bartoš 1948), S. Bohemia; *H. bulgarica* (Wisniewski 1933), high lakes in Caucasus, Andes & Himalayas; *H. fennica,* (Levander 1892), *H. jenkinae* (Beauchamp 1932), *H. libica* (Manfredi 1939), sea and salt water.

TYPE SPECIES: *H. mira* (Hudson 1871) (described as *Pedalion mira).*

For descriptions of various species see Bartoš (1948), Löffler (1954), Sládeček (1955), Wälikangas (1924), Wisniewski (1929).

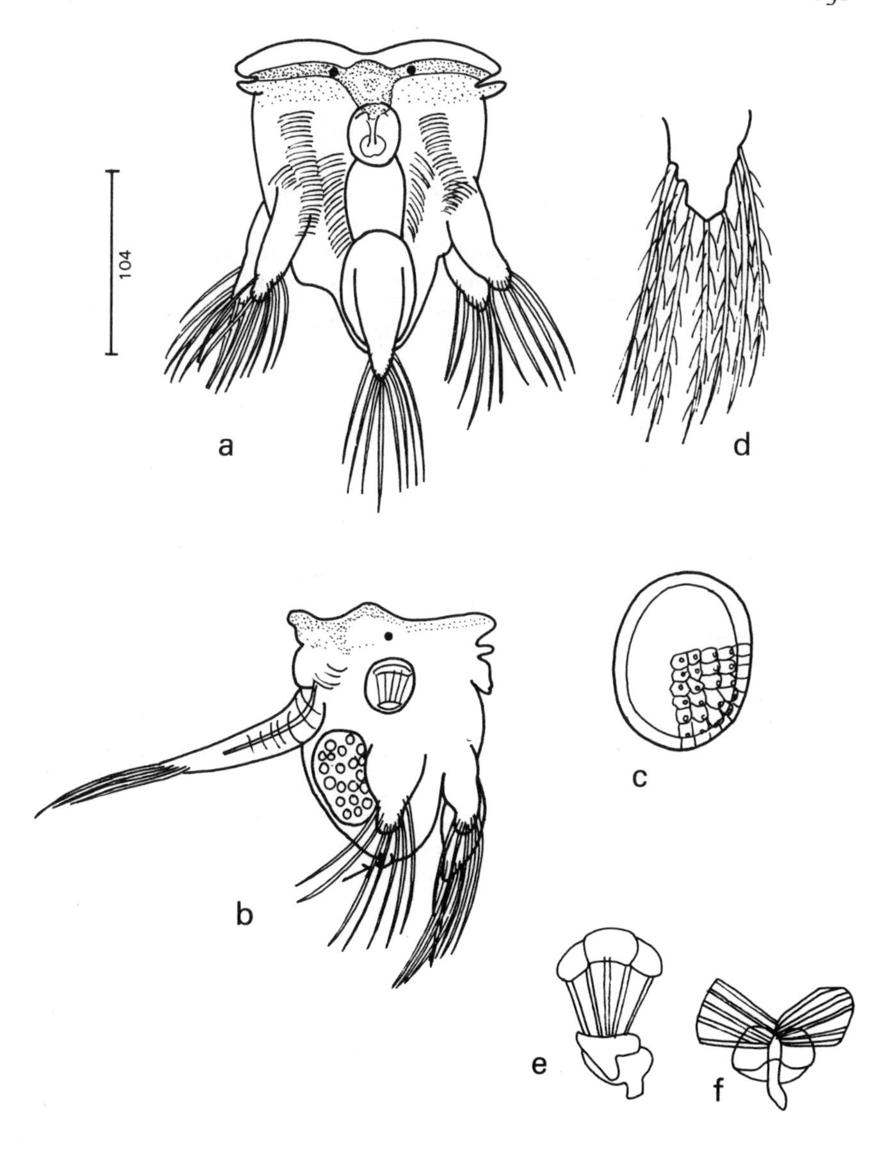

Fig. 111. *Hexarthra mira.* Female: *a*, dorsal; *b*, lateral, ↙ caudal process; *c*, sexual egg; *d*, lateral arm; *e*, jaws, lateral; *f*, jaws, dorsal.

Filinia Bory de St Vincent 1824

No foot; body cylindrical or sack-shaped, with 2 movable anterior lateral bristles and 1 posterior bristle (figs 112-116). Two red eyes on head. Resting eggs sink to bottom with dead mother; shell has rows of blisters which later fill with gas and aid rise of eggs. Males with ciliated apical process (fig. 126).

1 Length of lateral bristles considerably greater than body-length— **2**

— Length of lateral bristles only slightly greater than, or equal to or less than body-length— **4**

2 Posterior bristle arises terminally, i.e. at or within 10 µm of end of body (fig. 112). Body <190 µm; lateral bristles <470 µm, posterior bristle almost as long as laterals. Resting eggs Dec., Feb. and May—
(F. major) **Filinia terminalis** (Plate 1886)

Planktonic; lakes and ponds and brackish water; sometimes numerous.

— Posterior bristle arises from ventral surface, i.e. 15 µm or more from end of body (figs 113, 114)— **3**

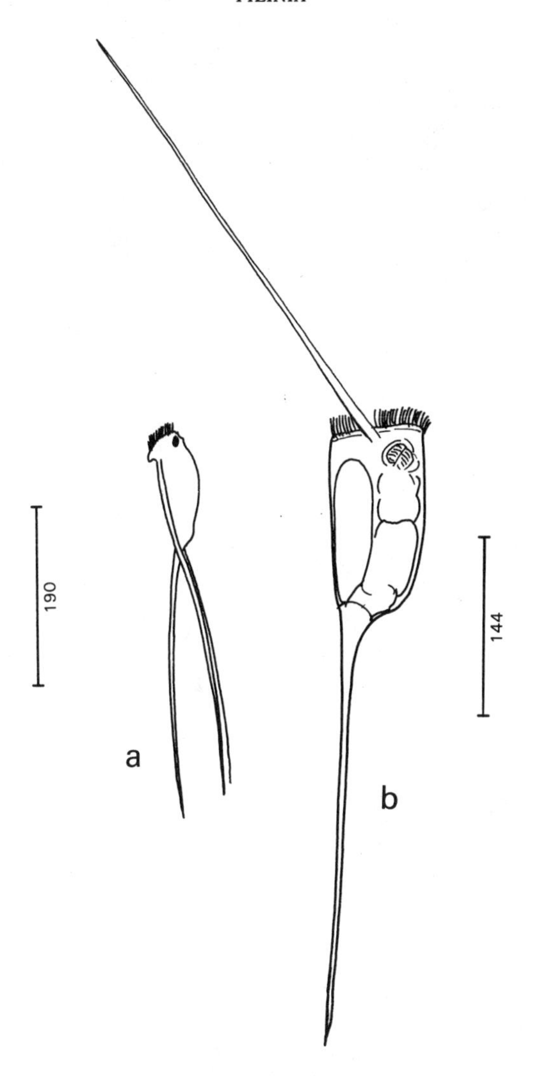

Fig. 112. *Filinia terminalis*. Female: *a*, young animal, lateral; *b*, older, lateral. (After Hollowday personal communication).

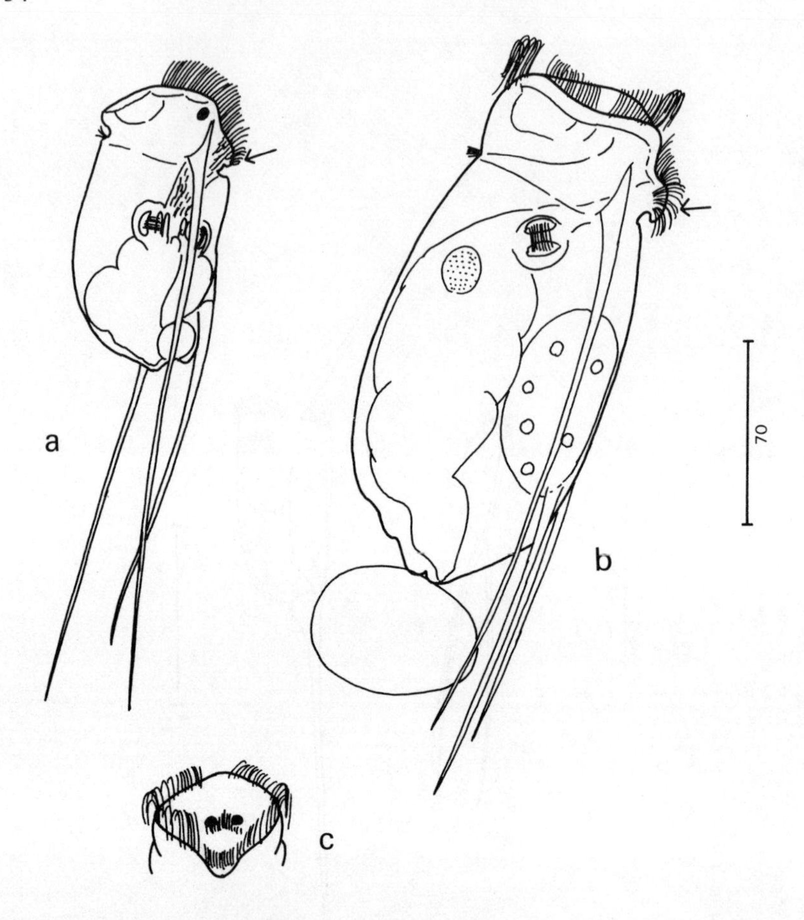

Fig. 113. *Filinia passa*. Female: *a*, newly-hatched, lateral, with relatively long bristles; *b*, fully-grown, bristles relatively shorter; *c*, ventral view of head; ↙ lip. (After Hollowday personal communication).

3 Lower 'lip' protrudes into a snout or beak (fig. 113*a*, *b* ↙). Length of bristles somewhat variable, but lateral bristles never reach twice body length in adult (young animals have longer bristles relative to body, like *F. longiseta*; as body grows longer, bristles remain about the same; lip also becomes more prominent with age. Fig. 113*a*, *b*). Body <200 µm. Resting egg with net-like shell; MALE <70 µm, April, May, Sept., Oct. (fig. 126*a*)— **Filinia passa** (Müller 1786)

Planktonic; ponds, canals.

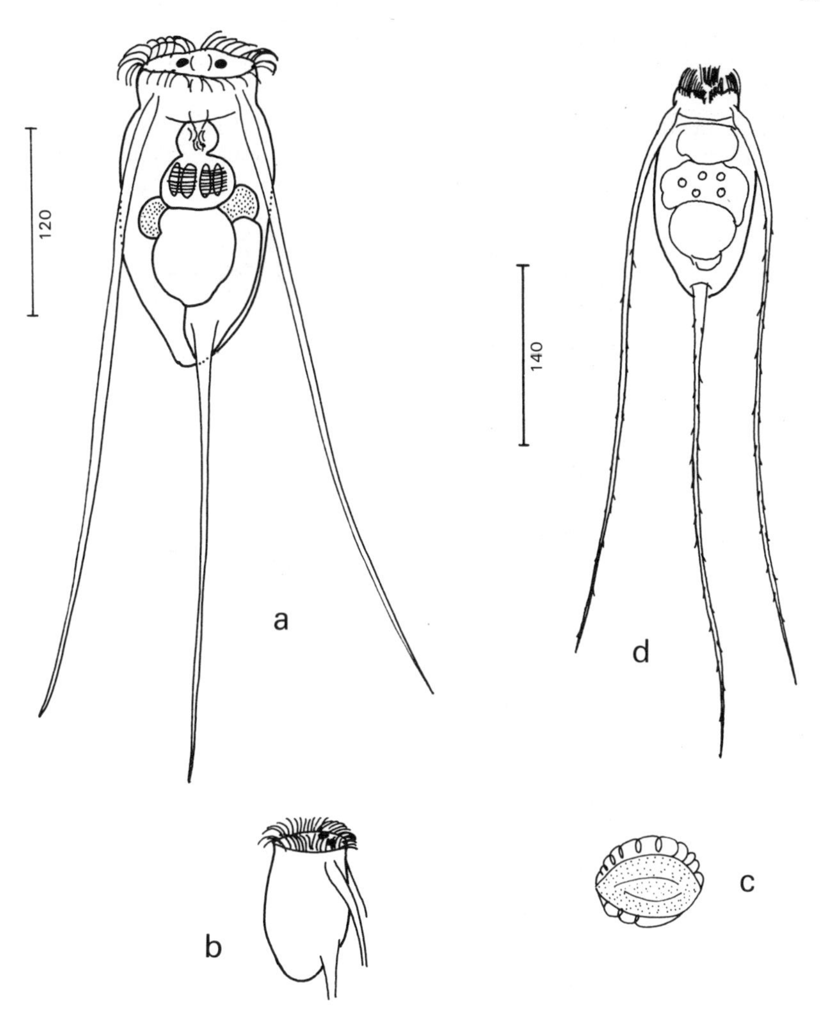

Fig. 114. *Filinia longiseta.* Female: *a,* ventral; *b,* lateral, base only of bristles shown; *c,* sexual egg; *d,* var. *limnetica,* ventral. (*d,* after Hollowday personal communication).

— Lower lip not protruded (fig. 114). Lateral bristles at least twice as long as body; length of body <250 μm, lateral bristles <510 μm, about twice as long as posterior bristle. Resting eggs spring and autumn; (fig. 114*c*). MALE <90 μm (fig. 126*c*)—

F. longiseta (Ehrb. 1834)

Cosmopolitan, planktonic; lakes, ponds, moor waters and brackish water; perennial species with one or two maxima each year; usually numerous; pond form usually with shorter bristles than lake form.

Form with lateral bristles more than 4 times as long as body; in large waters in summer (fig. 114*d*)— var. **limnetica** (Zacharias 1893)

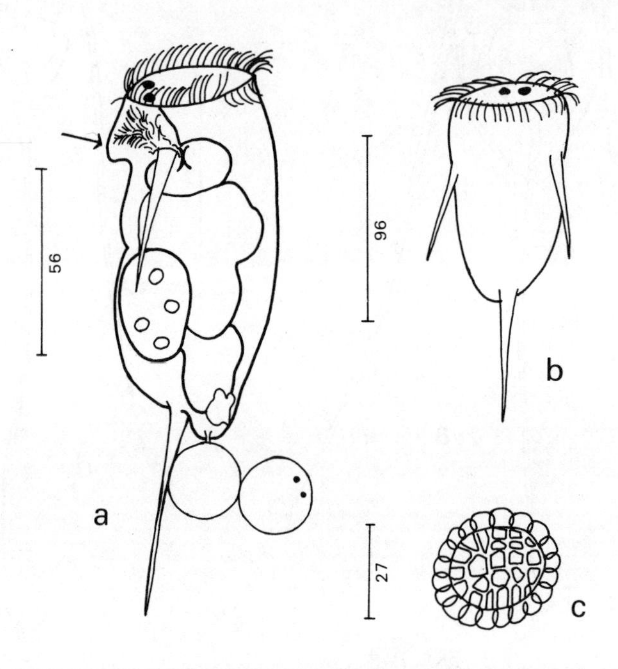

Fig. 115. *Filinia brachiata.* Female: *a*, lateral, with eggs, ↙ lip; *b*, ventral; *c*, sexual egg.

4 (1) Posterior bristle arises from ventral surface, i.e. 15 μm or more from end of body (fig. 115). Lower 'lip' protrudes clearly as snout or beak (fig. 115*a*↙). Bristles of variable length, with broad bases. Body <190 μm. MALE in June—

Filinia brachiata (Rousselet 1901)

Planktonic; ponds, pools, canals and brackish water, in summer.

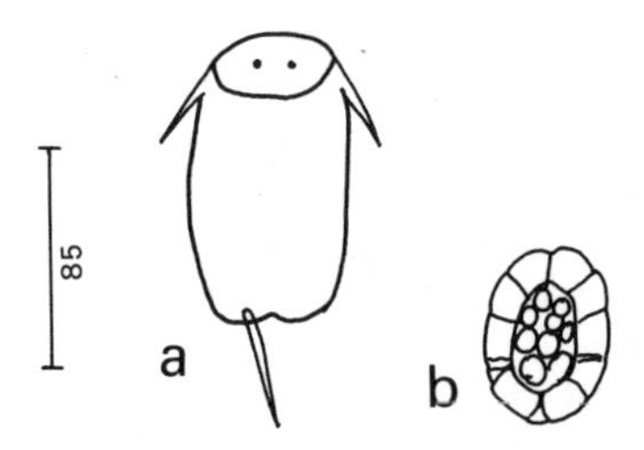

Fig. 116. *Filinia cornuta.* Female: *a,* ventral; *b,* sexual egg. (*a,* after Weisse 1847, *b,* after Dieffenbach & Sachse 1912).

— Posterior bristle arises terminally, i.e. at or within 10 µm of end of body (fig. 116). Lower lip protruded very little or not at all. Bristles about ½ body length. Body <120 µm. MALE <80 µm, May (fig. 126*b*)— **F. cornuta** (Weisse 1847)

Planktonic, widespread but sporadic in lakes, ponds, moor waters and brackish water in colder months.

OTHER SPECIES: *F. camascela* Myers 1938, Panama.

See also: *Tetramastix opoliensis* (Zacharias 1898) (?syn. *F. opoliensis),* tropical & warm waters; *Fadeewella minuta* Smirnov 1928 (?syn. *F. minuta),* Siberia. Both species resemble *Filinia,* but have 2 posterior bristles.

TYPE SPECIES: *F. passa* (Müller 1786) (described as *Brachionus passus*). Original description of genus in Bory de St Vincent (1824b).

For literature on *Filinia,* see Carlin (1943), Hutchinson (1964), Pejler (1957); on *Tetramastix,* see Albertova (1959); on *Fadeewella,* Smirnov (1928).

TROCHOSPHAERA Semper 1872

T. aequatorialis Semper 1872. Type species; found in the Philippines and China, in rice fields.

T. solstitialis Thorpe 1893, found in China, N. America, Africa, the Danube delta.

For further information see Rahm (1956) and Valkanov (1936).

HORAELLA Donner 1949

H. brehmi Donner 1949; found in plankton in a water tank in India, and a crater in Africa. For a description see Beadle (1963) and Donner (1949).

CONOCHILIDAE

CONOCHILUS Ehrb. 1834 & CONOCHILOIDES Hlava 1904

Lorica absent; foot stout, toes absent, no holdfast. Head convex, corona unlobed, horseshoe-shaped, with mouth in middle or near dorsal edge. Jaws malleoramate (fig. 7*b*). Transparent gelatinous case; solitary or in spherical colonies. (Figs 117-120). Planktonic.

1 Paired or fused ventral antennae situated below corona (fig. 117*a*); mouth in middle of corona (fig. 118↙). Case reaching about as far forward as ventral antennae. Dorsal antennae present (fig. 117*a*). Usually solitary. Asexual eggs laid in case; resting eggs with thin gelatinous case and a spiral structure (fig. 117*b*)—
CONOCHILOIDES, **2**

— Ventral antennae situated on coronal area (figs 119, 120); mouth near dorsal edge of corona (fig. 119*a*, *c*↙). Case round stalk. No dorsal antennae. Colonial, forming spherical colonies (which separate into component individuals in preserved matcrial) (fig. 120*a*). Resting eggs with thin gelatinous case and rupture ring (fig. 119)—
CONOCHILUS, **3**

2 Ventral antennae separate, long (fig. 117). Case surrounds plump foot and posterior end of body (fig. 117). Jaws symmetrical, 5-6 large teeth on uncus. Length <510 µm. MALE cone-shaped, <100 µm; resting eggs and males in April (fig. 117*b*)—
Conochiloides natans (Seligo 1900)

Cosmopolitan; cold stenotherm, winter and spring; sometimes numerous in lakes and ponds.

— Ventral antennae fused (fig. 118). Foot more slender. Jaws somewhat asymmetrical, with 3 large teeth on one uncus, 5 on the other. Length <500 µm. Males unknown—
Conochiloides dossuarius (Hudson 1885)

Cosmopolitan, planktonic; lakes, moor waters, shallow ponds; warmer waters.

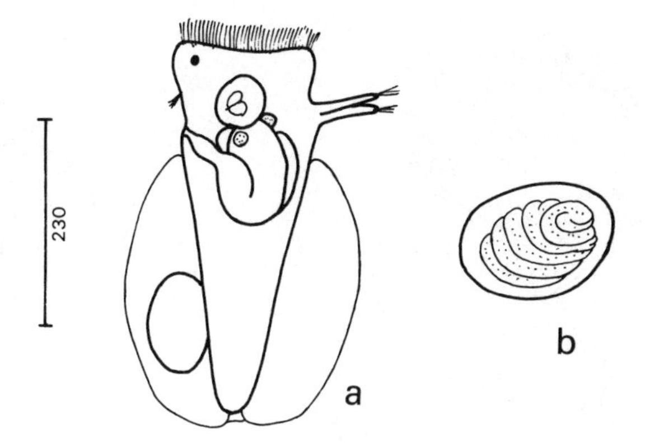

Fig. 117. *Conochiloides natans.* Female: *a,* lateral; *b,* sexual egg. (*a,* after Voigt 1904, *b,* after Ruttner-Kolisko 1974).

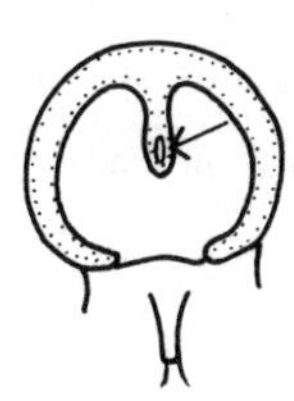

Fig. 118. *Conochiloides dossuarius.* Female: position of mouth and ventral antennae, ↙ mouth. (After Ruttner-Kolisko 1974).

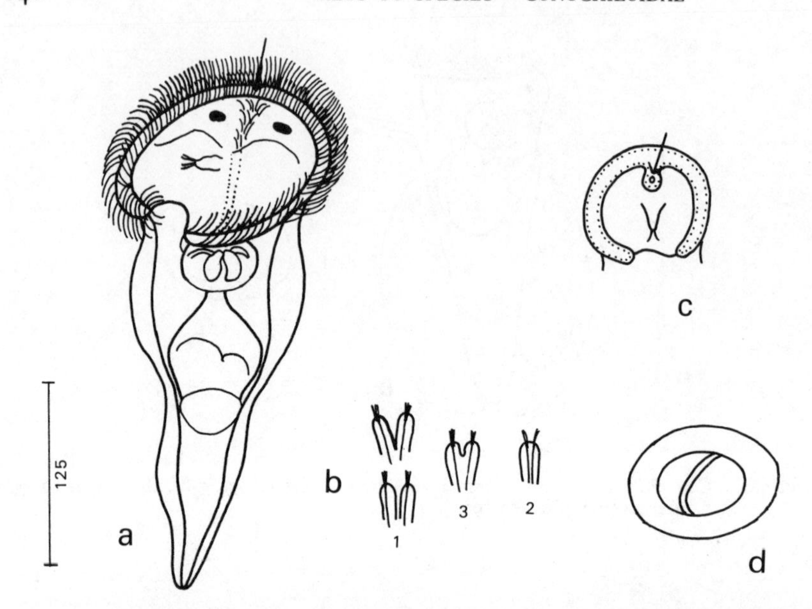

Fig. 119. *Conochilus.* Female: *a,* individual of *C, unicornis,* ventral, ↙ position of mouth; *b,* forms of ventral antennae – 1, typical *C. hippocrepis,* 2, typical *C. unicornis,* 3, intermediate form of antennae; *c,* position of mouth and ventral antennae in *C. unicornis,* ↙ mouth; *d,* sexual egg. (*c, d,* after Ruttner-Kolisko 1974).

3 (1) Colony of about 5-12 individuals. Ventral antennae fused (fig. 119*a, c*). Length <380 µm; colonies <1 mm diameter. Extended foot about as long as body. Perennial species, maximum spring/summer, resting eggs June-Nov. (fig. 119*d*)—
Conochilus unicornis (Rousselet 1892)

Cosmopolitan; lakes and ponds, and brackish water.

— Colony of about 30-60 individuals (fig. 120*a*). Ventral antennae separate or only partly fused (fig. 120*b, c*). Length <800 µm; colonies <4 mm diameter. Extended foot about twice as long as body (fig. 120c). Maximum April, May. (Male fig. 127)—
Conochilus hippocrepis (Schrank 1803)

Cosmopolitan; lakes, ponds, moor waters.

Considerable variation occurs in *Conochilus,* with colonies intermediate in some respects between *C. unicornis* and *C. hippocrepis.* In particular, forms of ventral antennae intermediate between the completely separate and completely fused occur (fig. 119*b*). There is also variability in size of colony and length of foot.

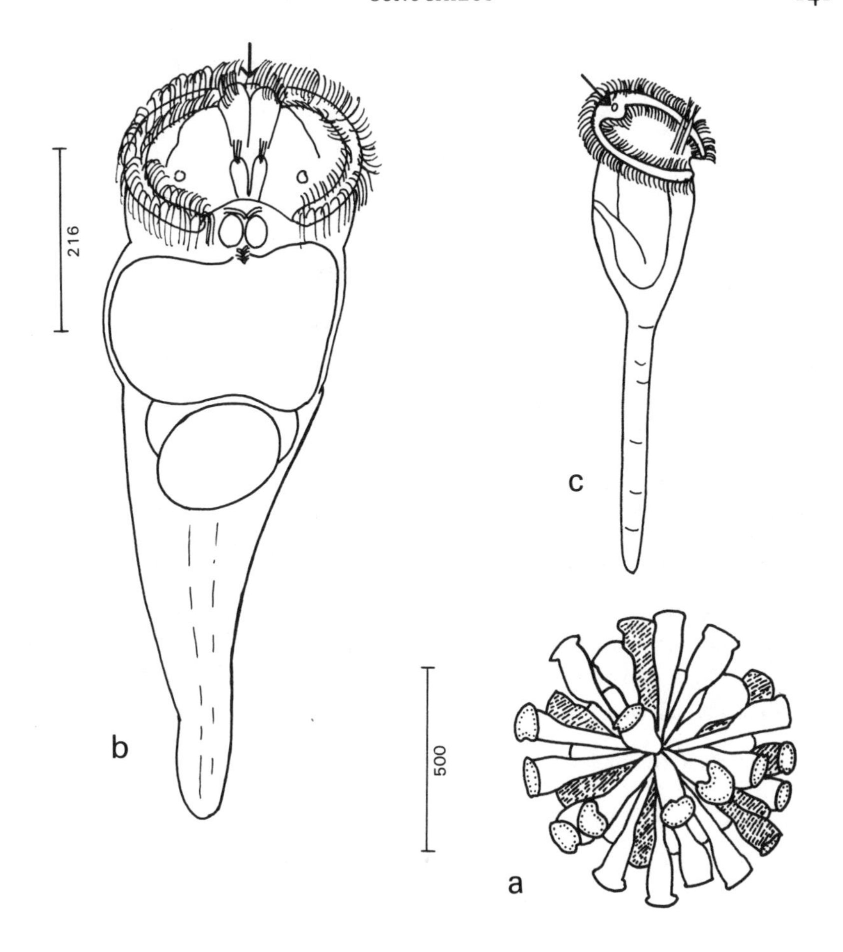

Fig. 120. *Conochilus hippocrepis.* Female: *a,* colony; *b,* individual, ventral; *c,* lateral, ↙ position of mouth. (*c,* after Bartoš 1959).

OTHER SPECIES: (both planktonic) *Conochiloides coenobasis* Skorikov 1914, Russia, Sweden, N. America; *Conochiloides exiguus* Ahlstrøm 1938, N. America.

These 2 species have tooth number on uncus of 3+3 for *C. coenobasis,* 4+5 for *C. exiguus.*

TYPE SPECIES: *Conochilus hippocrepis* (Schrank 1803) (described as *Linza hippocrepis); Conochiloides natans* (Seligo 1900) (described as *Tubicolaria natans).*

For information on the genera see Hlava (1905), Pejler (1957), Pourriot (1965); on colony building, see Ruttner-Kolisko (1939).

COLLOTHECIDAE

Collotheca Harring 1913

Lorica absent; foot stalk-like; toes absent. Transparent gelatinous case usually present. Head funnel-like with central mouth; corona circular or with 1, 2 or 5 lobes often bearing very long cilia (figs 121-123). Jaws uncinate (fig. 6). Mostly sedentary, solitary individuals; a few planktonic species.

1 Corona with 2 lobes (fig. 121); lobes not always easy to distinguish, but cilia of lobes longer than those of coronal border between lobes. Case wide, somewhat irregular, with rounded end. Foot usually with spindle-shaped thickening and with holdfast (fig. 121*d*). Two red eyes, wide apart, sometimes with red pigment spots on coronal area. Length $<580\ \mu m$. Usually found in summer; resting eggs in autumn. (Male fig. 145*a*)—
Collotheca mutabilis (Hudson 1885)

Cosmopolitan, planktonic; lakes and ponds, and brackish water.

— Corona with one lobe or unlobed (figs 122, 123)— **2**

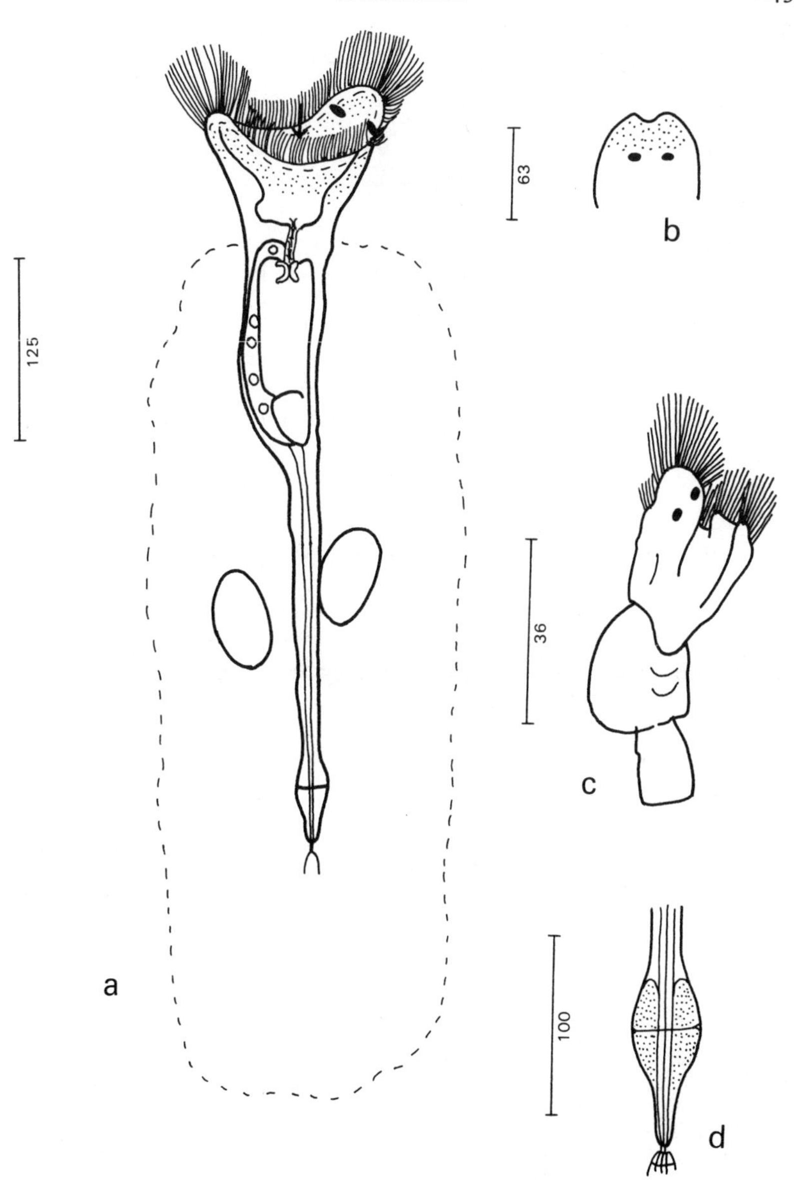

Fig. 121. *Collotheca mutabilis.* Female: *a*, extended, in case, with eggs, lateral, ↙ mouth; *b*, head with retracted corona, dorsal; *c*, young; *d*, end of stalk.

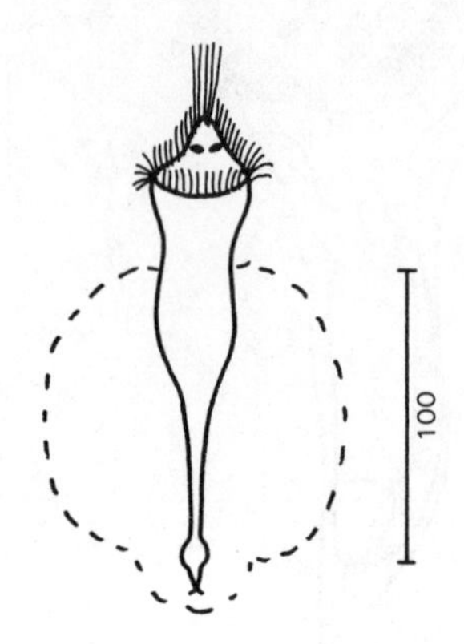

Fig. 122. *Collotheca libera.* Female: ventral. (After Voigt 1957).

2 Corona with one lobe, bearing long cilia (fig. 122). Case end rounded or pointed. Foot usually with onion-shaped thickening (fig. 122). Two eyes close together, with lenses. Males unknown—
Collotheca libera (Zacharias 1894)

Widespread but sporadic; July-Oct.; lakes and ponds

— Corona unlobed, circular, although long cilia occur in 5 groups, each group centred on a thickening or rudimentary lobe on inner side of coronal border (fig. 123↙). Foot rod-shaped, sometimes thickened. Eyes absent in older animals. Length <500 µm. Mar.-Dec., resting eggs Sept. (Male fig. 145*b*)—
C. pelagica (Rousselet 1893)

Cosmopolitan; lakes, ponds, moor waters and brackish water.

OTHER SPECIES: planktonic, not recorded for Britain and Ireland: *C. balatonica* Varga 1936, Bohemia. *C. lie-pettersoni* Bērziņś 1951. *C. tubiformis* Nipkow 1961, L. Zurich. *C. discophora* Skorikov 1903. *C. ornata-natans* Tschugunov 1921, Caspian Sea. *C. polyphema* Harring 1914, America. *C. undulata* Sládeček 1968.

TYPE SPECIES: *C. campanulata* (Dobie 1849) (described as *Floscularia campanulata).* Original mention of genus in Harring (1913b).

For information on and key to the many sedentary species, see Voigt (1957), also Bērziņś (1951); for description of *C. balatonica,* see Varga (1936), and of *C. undulata,* see Sládeček (1968).

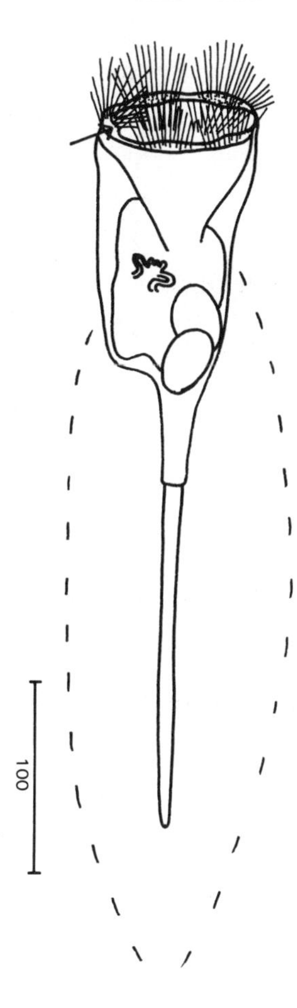

Fig. 123. *Collotheca pelagica*. Female: lateral, ↙ coronal thickening or rudimentary lobe.

PHILODINIDAE

ROTARIA Scopoli 1777

Lorica absent; foot present; 3 toes and 2 spurs on foot (fig. 124). Capable of leech-like creeping by telescoping of body and foot sections. Front of head forms snout or rostrum which takes the lead during creeping. Corona of 2 circles of cilia, side by side, used for swimming and feeding, retracted during creeping. Jaws ramate (fig. 7*a*). Two red eyes on snout. Males and sexual eggs unknown. Most species are found in mud, moss etc., but some are also free-swimming and occasionally occur in the plankton.

Foot very long and thin (fig. 124*b*)—

Rotaria neptunia (Ehrb. 1832)

Widespread but sporadic in muddy waters, canals; between plants, in detritus, on *Nepa,* and in free water.

TYPE SPECIES: *R. rotatoria* (Pallas 1766) (described as *Brachionus rotatorius). R. rotatoria* has occasionally been recorded from plankton.

For a key to, and descriptions of, the non-planktonic species see Voigt (1957) and Donner (1965).

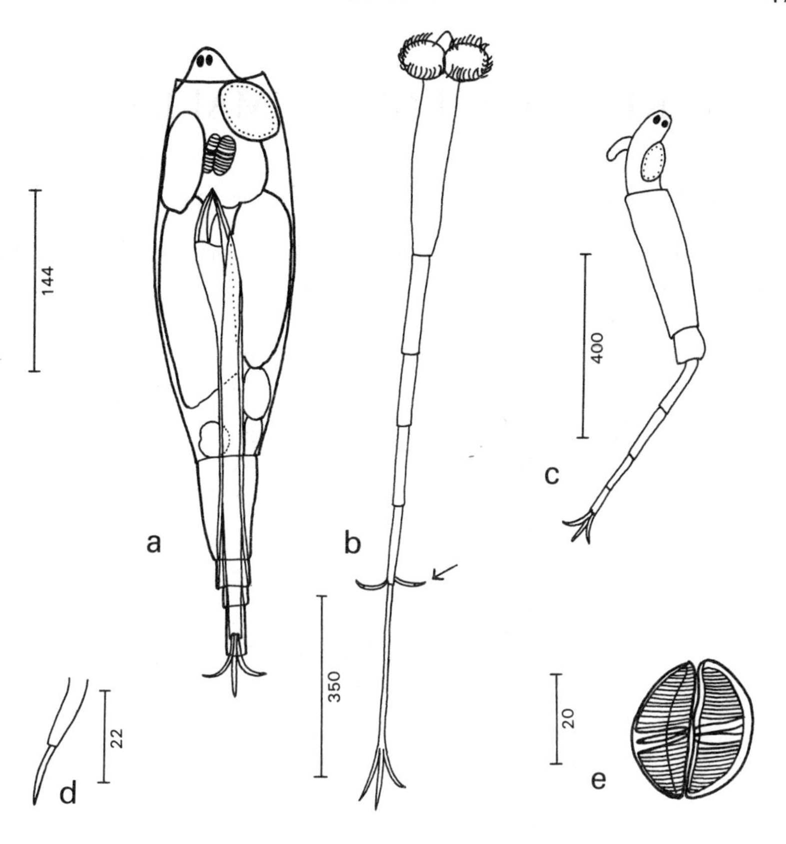

Fig. 124. *Rotaria neptunia.* Female: *a*, corona and foot retracted, lateral; *b*, extended, ventral, ↙ spur; *c*, creeping, head and foot extended, corona retracted, lateral; *d*, spur; *e*, jaws.

KEY TO GENERA (MALES)

1 Apical process or snout present, with 2 eyes (figs 125-127)— **2**

— Apical process absent; eyes, where present, on brain or head (e.g. figs 128-130)— **4**

2 Gut and jaws complete, similar to female, probably functional (fig. 125). Foot long, with 2 long foot-glands; body shaped like female and *c.* $\frac{2}{3}$ of its size, soft, transparent (fig. 125, female fig. 39). Corona well-developed; long snout as in female with 2 red eyes. Large testis, 2 pairs of prostate glands. Ciliated eversible sperm duct acting as copulatory organ. Length <360 µm. Spring— RHINOGLENA

Female, p. 46.

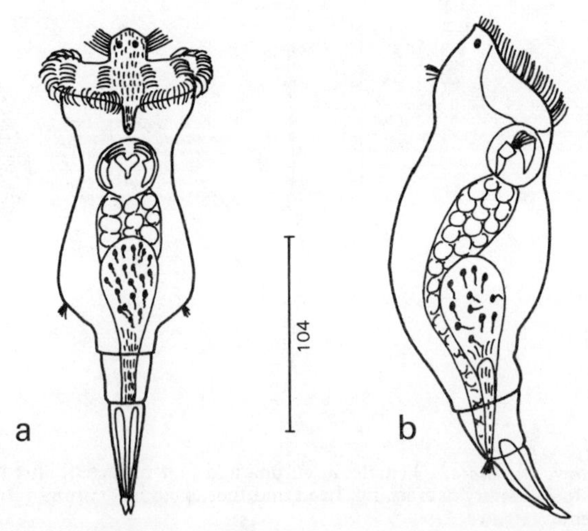

Fig. 125. *Rhinoglena frontalis.* Male: *a,* ventral; *b,* lateral. (After Wesenberg-Lund 1923).

— No trace of gut present (figs 126, 127). Small, conical to elongate, soft; no foot, body terminating in circle of cilia. Apical process conical, ciliated, with 2 red eyes. Testis large, pear-shaped. Oil globules in body cavity— **3**

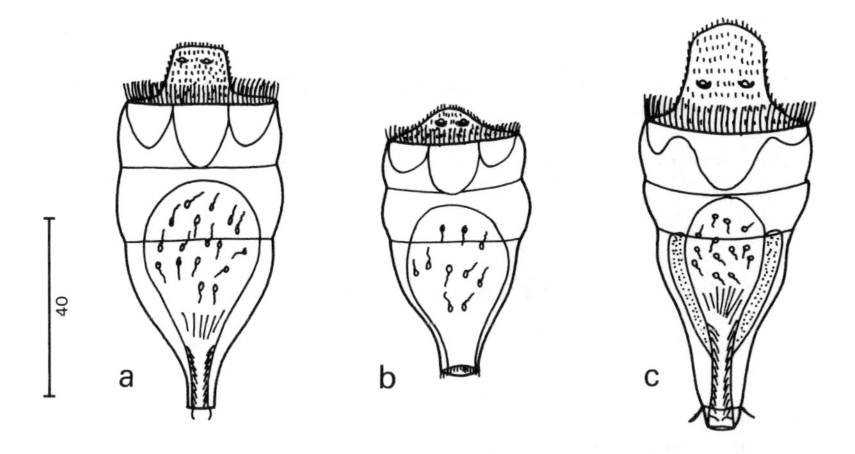

Fig. 126. *Filinia.* Male: *a, F. passa; b, F. cornuta; c. F. longiseta.* (After Wesenberg-Lund 1923).

3 Drawn-out end of body forming copulatory organ; 2 prostate glands (fig. 126). Length <90 μm. May— FILINIA

Female, p. 132.

— Ciliated sperm duct everted for copulation. No foot, but foot-glands (or prostate glands) present (fig. 127). Length <50 μm. April— CONOCHILUS

Female, p. 138.

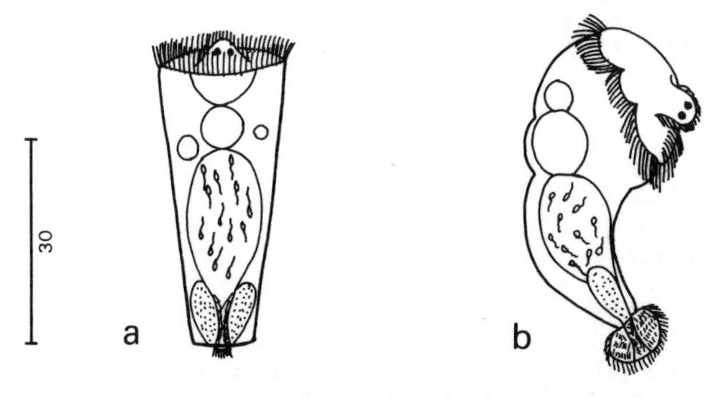

Fig. 127. *Conochilus hippocrepis.* Male: *a,* dorsal; *b,* lateral. (After Wesenberg-Lund 1923).

4 (1) Foot and penis both present and conspicuous; or penis only present and conspicuous, no foot (NOTE: appearance of penis may be judged somewhat subjectively; see figs 128-133. Foot and penis may be retractile)— **5**

— Foot present or absent, penis short and inconspicuous or absent (e.g. figs 134, 135, 139)— **10**

5 Both foot and penis present and conspicuous (figs 128, 129)— **6**

— Penis present and conspicuous, foot absent (figs 130-133)— **7**

6 Foot ending in 2 short conical toes (fig. 128); foot probably not retractile. Body small, strongly reduced; lorica present, of dorsal and ventral plates, delicate, no spines. Corona retractile. Red eye on brain. Gut rudiment a strand supporting testis, which is large with prostate glands. Penis retractile, very large when extended, thick, wrinkled, tapering, with long cilia; traversed by sperm duct which is ciliated anteriorly, and posteriorly ends in stiff tube with opening in disc-shaped apex (fig. 128). Length <150 µm— BRACHIONUS

Males well-known: *B. calyciflorus* <120 µm, lorica faintly developed; *B. rubens* <130 µm; *B. angularis* <90 µm, characteristically shaped lorica, no trace of gut (fig. 128*a-i*). Female, p. 50.

— Foot long, stout, retractile, no toes, terminates in circle of cilia, as in female (fig. 129; female fig. 108). Body not flattened as female but somewhat elongated; lorica present. Two red eyes. Gut rudiment present. Testis and penis large (fig. 129). Length <135 µm. May— TESTUDINELLA

Female, p. 126.

Fig. 128. *Brachionus*. Male: *a-c*, *B. angularis*, *a*, dorsal; *b*, ventral; *c*, lateral; *d*, *B. quadridentatus*, dorsal; *e*, *B. rubens*, lateral; *f*. *B. leydigi*; *g*, *h*, *B. calyciflorus*, *g*, dorsal; *h*, lateral; *i*, *B. urceolaris*. (*e*, *f*, after Rousselet 1907, *g-i*, after Wesenberg-Lund 1923).

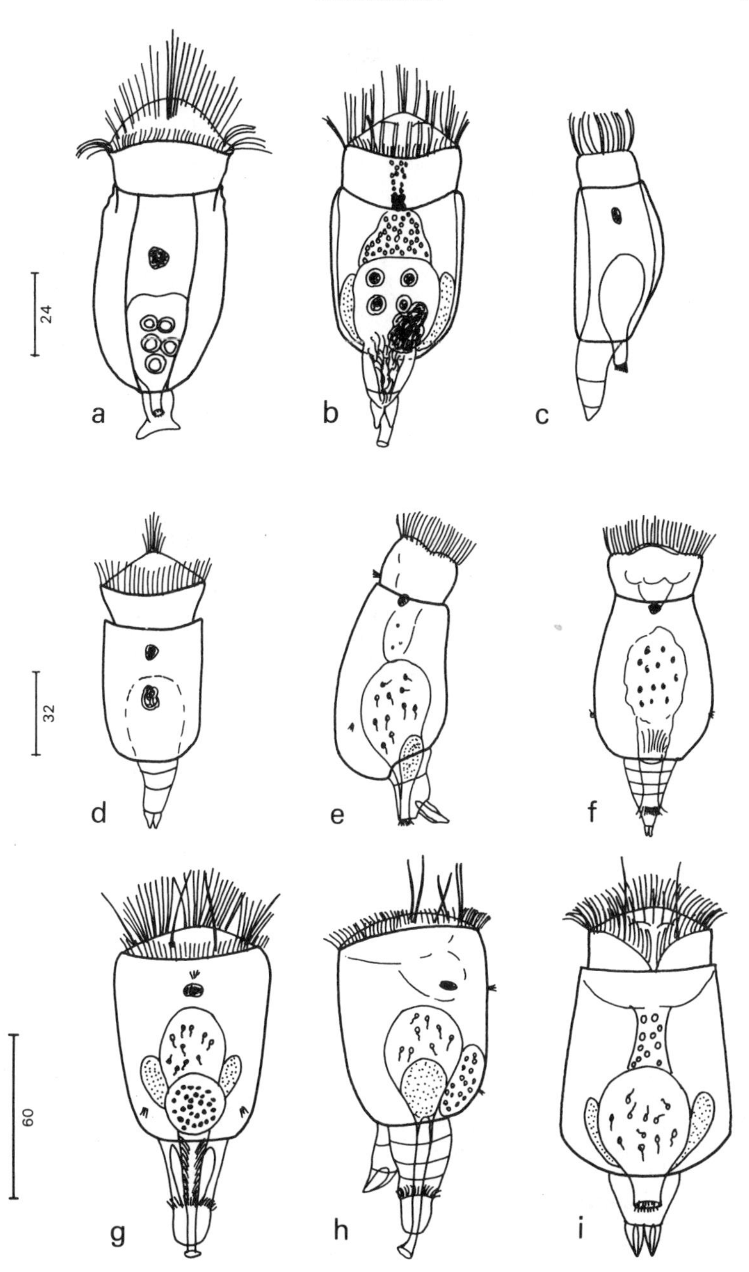
24
a
b
c
32
d
e
f
60
g
h
i

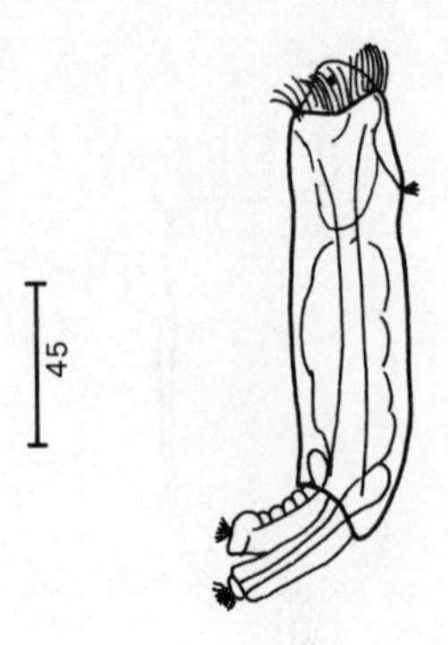

Fig. 129. *Testudinella patina.* Male, lateral. (After Wesenberg-Lund 1923).

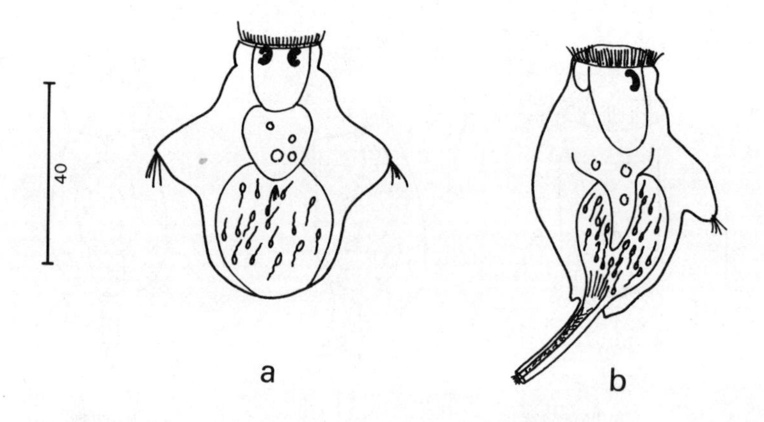

Fig. 130. *Hexarthra mira.* Male: *a*, dorsal; *b*, lateral. (After Wesenberg-Lund 1923).

7 (5) Three arms or appendages present, 1 dorsal and 2 lateral, terminating in bunches of cilia (fig. 130). Body broad to globular. Corona a ciliated disc; large brain with 2 curved red eyes. Testis large, globular. Penis very long, $<\frac{1}{2}$ body length, retractile in swimming, containing ciliated sperm duct (fig. 130*b*). Length $<$50 µm. Aug.— HEXARTHRA

Female, p. 130.

— No arms or appendages (figs 131-133)— **8**

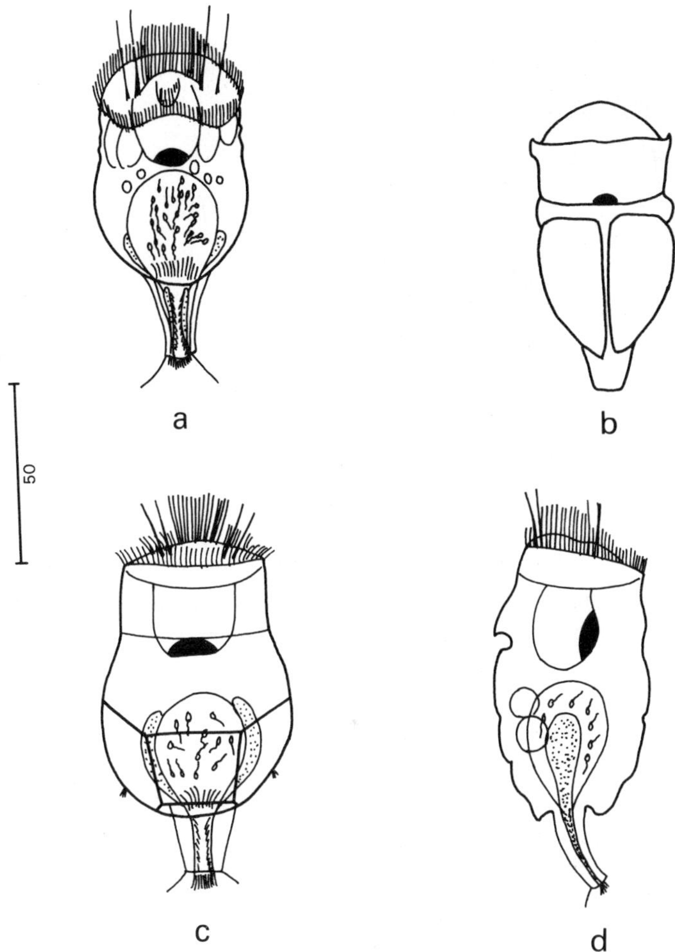

Fig. 131. *Keratella.* Male: *a, b, K. cochlearis, a,* dorsal; *b,* lorica; *c, d, K. quadrata, c,* dorsal; *d,* lateral. (After Wesenberg-Lund 1923).

8 Lorica partially developed, cuticle fairly thick, delicate dorsal plate present (fig. 131). Body small, conical. Large brain with red eye. Gut absent. Large testis, surrounded by oil drops and a conspicuous globule. Penis well-developed; sperm duct ciliated, prostate glands present (fig. 131)— KERATELLA & KELLICOTTIA

Keratella quadrata <100 µm, penis 20 µm, 3-part dorsal plate; *Keratella cochlearis* <90 µm, penis 20 µm (fig. 131*a-d*). Female, p. 66.

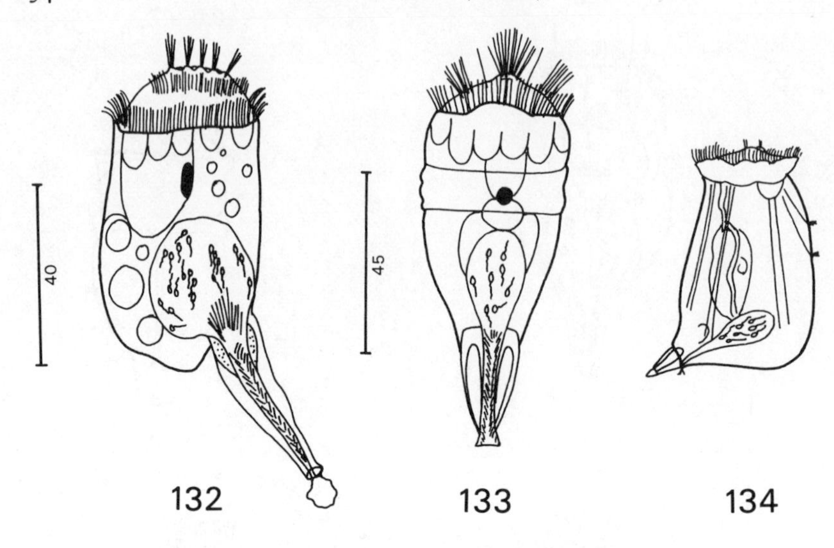

Fig. 132. *Notholca acuminata*. Male: lateral. (After Wesenberg-Lund 1923).

Fig. 133. *Anuraeopsis fissa*. Male: dorsal. (After Wesenberg-Lund 1923).

Fig. 134. *Asplanchnopus multiceps*. Male: lateral. (After Voigt 1957).

— No trace of lorica (figs 132, 133)— **9**

9 Body elongated, very soft, transparent (fig. 132). Large brain with red eye. Gut absent; large oil globules in body cavity. Large testis, 2 prostate glands. Penis very long, nearly as long as body, tapering to apex, flexible; sperm duct ciliated (fig. 132)— NOTHOLCA

Female, p. 75.

— Body elongated, soft, transparent, with 1 ventral and 2 dorsal furrows (fig. 133). Large brain with red eye. Gut absent; large oil globule near brain. Small testis, prostate glands. Penis very large and long, soft and flexible; sperm duct ciliated, stiffened, ending in cup-shaped disc (fig. 133). Length <90 μm; large relative to female; large penis, <30 μm, used as rudder in swimming— ANURAEOPSIS

Female, p. 64.

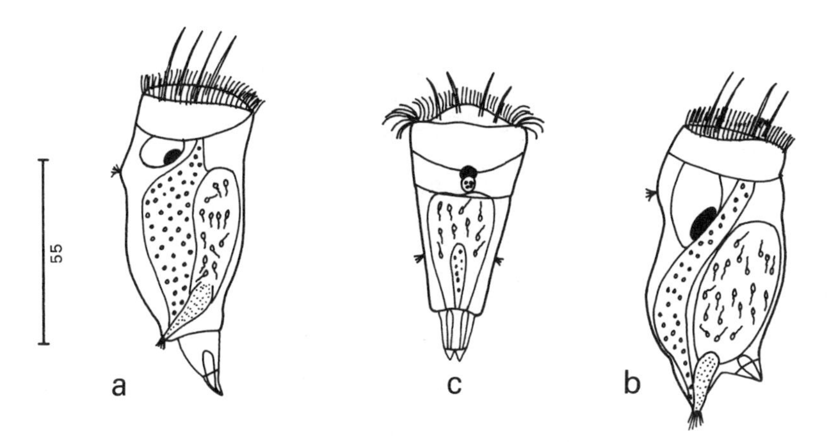

Fig. 135. *Synchaeta.* Male: *a, S. pectinata,* lateral; *b, S. tremula,* lateral; *c, S. oblonga,* dorsal. (*a, b,* after Wesenberg-Lund 1923, *c,* after Rousselet 1902).

10 (4) Foot present— **11**

— Foot absent or uncertain — difficulty arises in some genera in deciding whether body terminates in bunch of cilia, or a definite foot is present (figs 139-145)— **15**

11 Gut almost complete, but non-functional (fig. 134). Contractile vesicle present, large, many flame bulbs each side. Foot as female, fig. 86 (and p. 100). Penis eversible— ASPLANCHNOPUS

— Gut rudimentary (figs 135-138). Contractile vesicle absent— **12**

12 Four stiff bristles or styli on head (fig. 135) as in female, fig. 101. Body conical, cuticle soft and transparent. Foot short; 2 short toes; 2 well-developed foot-glands. Corona a simple circle of cilia, no auricles as female has. Large brain with large red eye. Gut rudiment a sac of oil globules. Testis large, pear-shaped; 2 short prostate glands. Penis short, dorsal to foot, rounded, cilia at opening (fig. 135). Length <160 µm— SYNCHAETA

Males similar in *S. pectinata, tremula, oblonga* (fig. 135*a-c*); in many species males unknown. Female, p. 116.

— No styli on head (figs 136-138)— **13**

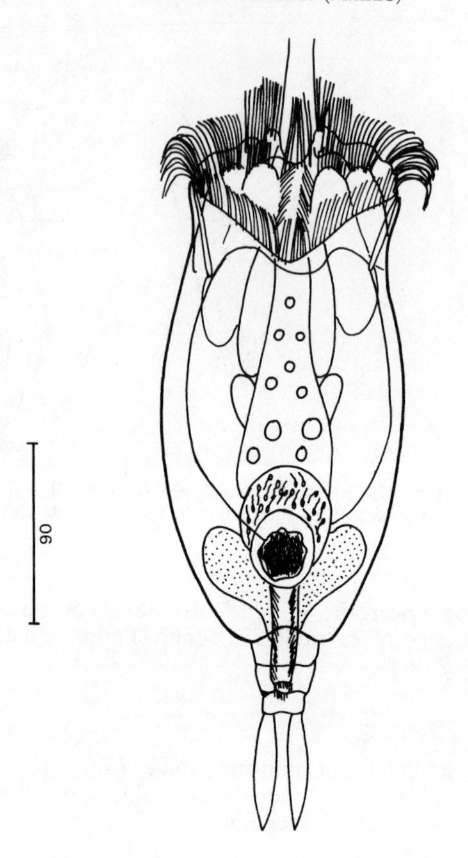

Fig. 136. *Euchlanis dilatata*. Male: ventral. (After Wesenberg-Lund 1923).

13 Toes large, leaf-shaped (fig. 136). Body shape similar to female (fig. 48). Lorica present, like that of female; very transparent. Foot 2- or 3-sectioned, although divisions not always clear; foot-glands present or absent. Corona well-developed as in female. Very large retrocerebral organ behind brain; red eye. Gut rudiment a strand supporting testis. Testis large; 2 prostate glands. Penis short, ciliated, protrusible (fig. 136). Length $\frac{1}{2}$-$\frac{2}{3}$ female— EUCHLANIS

E. dilatata <320 µm. April-June. Female, p. 61.

— Toes small or absent (figs 137, 138)— **14**

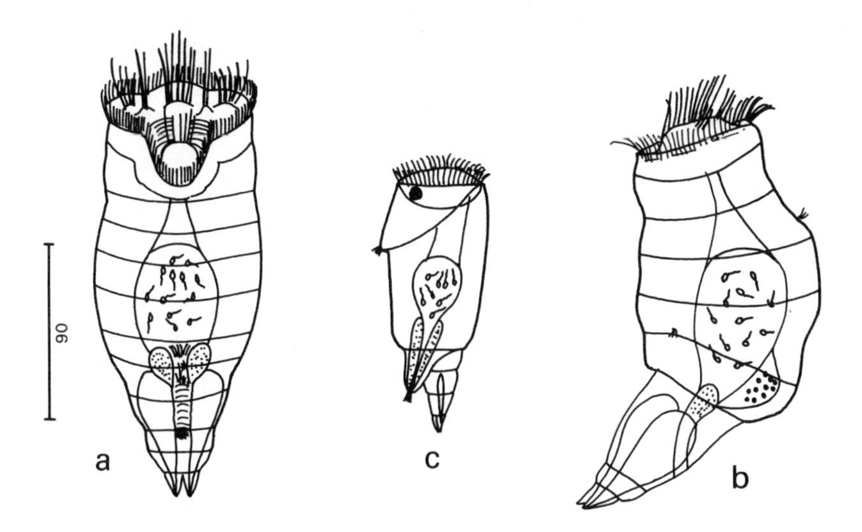

Fig. 137. *Epiphanes*. Male: *a, b, E. senta, a,* ventral; *b,* lateral; *c, E. brachionus,* lateral. (*a, b,* after Wesenberg-Lund 1923, *c,* after Weber 1898).

14 Foot short to long, 2 short toes, 2 large foot-glands (fig. 137). Body similar to female (fig. 41), cuticle thin and transparent. Corona well-differentiated; eye present or absent. Gut rudiment a strand supporting large testis, 2 prostate glands, eversible penis— EPIPHANES

E. senta <250 µm, March-May, no eye. *E. brachionus* <160 µm, red eye (fig. 137*a-c*). Female, p. 48.

— Foot short, 2 short toes. May have 'bubbly' lorica and 2 palps on head. Oil globules forming granular mass filling body; testis small, protrusible or eversible penis (fig. 138)— PLOESOMA
Female, p. 122.

Note. *Trichocerca* may key out here, as foot may be present or absent. See couplet 16.

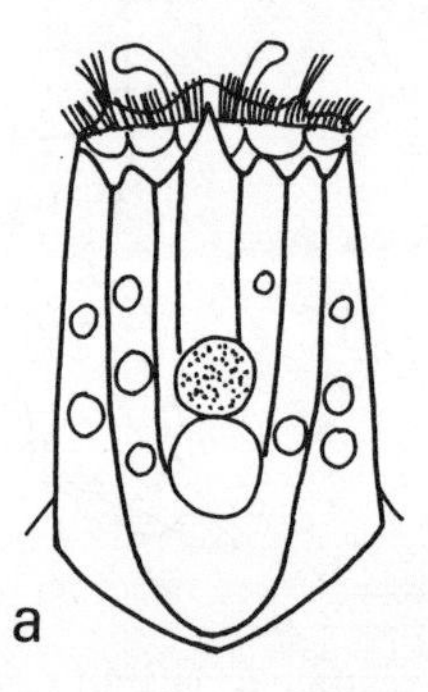

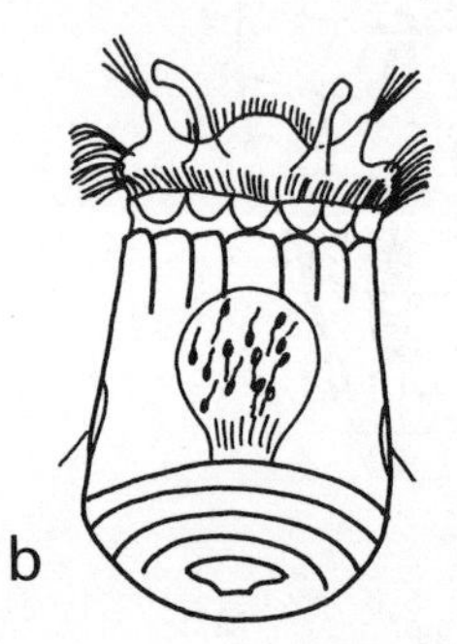

Fig. 138. *Ploesoma hudsoni.* Male: *a,* dorsal; *b,* ventral. (After Wesenberg-Lund 1923).

15(10) Contractile vesicle present, large (fig. 139). Body little reduced, similar to female (fig. 88), very transparent. Corona a simple circle of cilia. One or more red eyes. Some rudiment of gut present. Copulatory organ formed by eversion of sperm duct (fig. 139*c*)— ASPLANCHNA

A. priodonta similar in shape to female but narrower; gut rudiment a strand holding testis and a globular mass representing jaws; 1 large eye and 2 smaller; 3-4 flame bulbs on each side; testis pear-shaped. Length <500 μm (fig. 139*a*). *A. brightwelli* group globular or conical with humps; gut rudiment a globular mass in body cavity; single eye; 10-50 flame bulbs on each side; testis round. Length <1200 μm (fig. 139*b, c*). Female, p. 102.

— Contractile vesicle absent (figs 140-145)— **16**

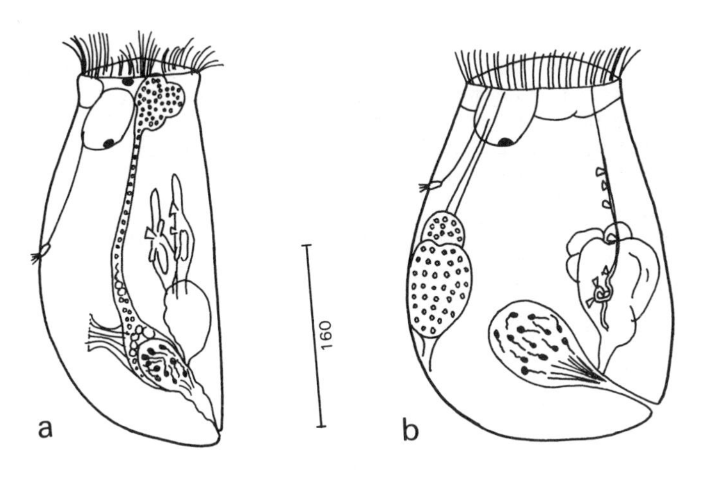

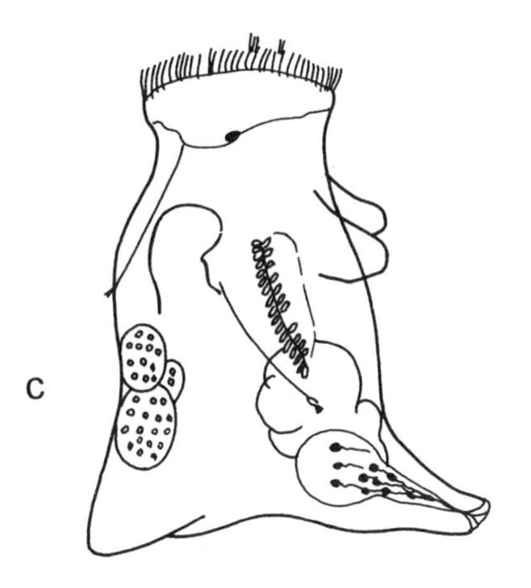

Fig. 139. *Asplanchna*. Male: *a*, *A. priodonta*, lateral; *b*, *A. brightwelli*, lateral; *c*, *A. sieboldi*, lateral. (*c*, after Hudson & Gosse 1886).

16 Testis large and globular; sperm duct long, thin, retractile, stiffened, ending in disc with opening; penis eversible (fig. 140). Body small, $\frac{1}{4}$-$\frac{1}{3}$ size of female, pear-shaped. Soft cuticle marked with lines and folds, especially over retracted corona; no crests or spines. Foot present or absent. Eye present or absent. No gut; many oil drops round testis. Body and organs symmetrical— TRICHOCERCA

Male known only in some species. *T. cylindrica* has cuticle marked with lines as of a lorica; no eye; 2 prostate glands, large oil globule and smaller drops; length <80 µm (fig. 140). *T. pusilla* has cuticle marked with transverse lines; red eye; length <60 µm. Female, p. 86.

— Sperm duct not stiffened as above— **17**

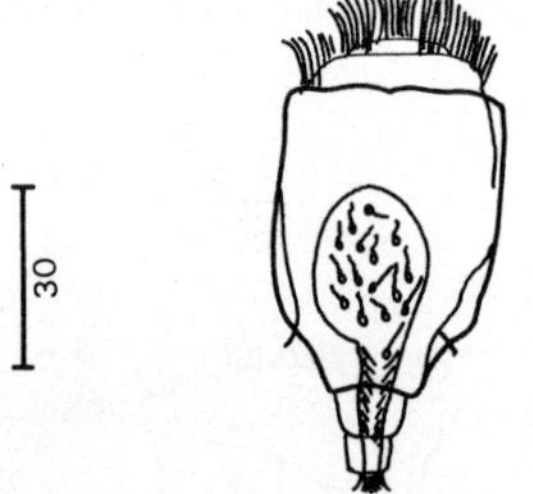

Fig. 140. *Trichocerca cylindrica.* Male: dorsal. (After Sudzuki 1956).

17 Eversible penis present— **18**

— Penis absent— **19**

18 Three genera, difficult to separate on incomplete information, key out here:

18(a) Delicate lorica may be present (fig. 141). Eye present or absent. Gut may be absent or a plasma strand. Testis large with prostate glands; ciliated sperm duct. Length <80 µm— GASTROPUS

G. hyptopus: delicate lorica, folds along dorsal side; red eye; testis globular (fig. 141*a*, *b*); May. *G. stylifer:* lorica rather thick; no eye; pear-shaped testis; Aug. (fig. 141*c*, *d*). Female, p. 95.

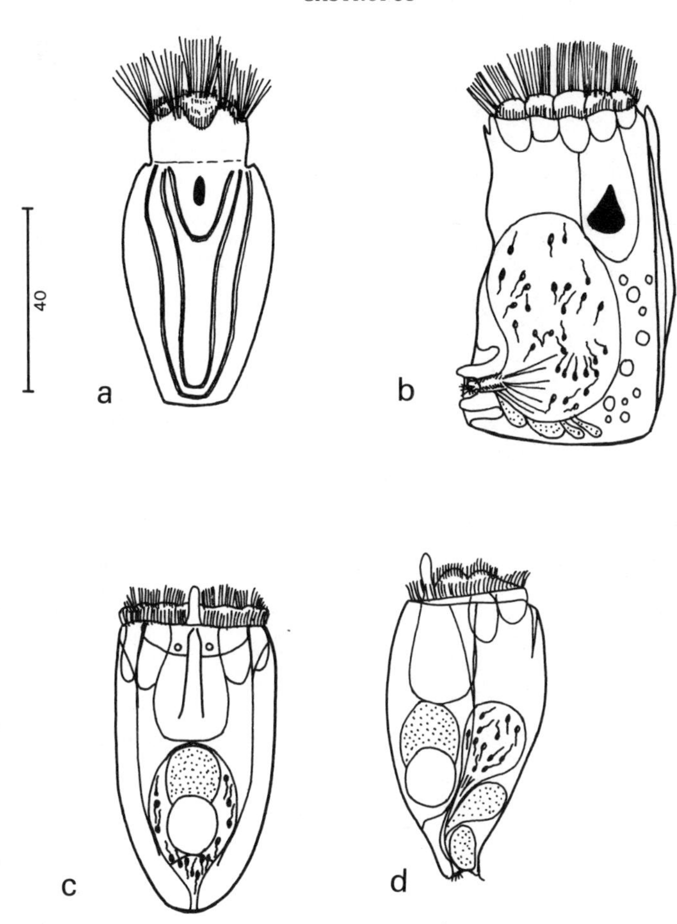

Fig. 141. *Gastropus*. Male: *a*, *b*, *G. hyptopus*, *a*, dorsal; *b*, lateral; *c*, *d*, *G. stylifer*, *c*, dorsal; *d*, lateral. (After Wesenberg-Lund 1923).

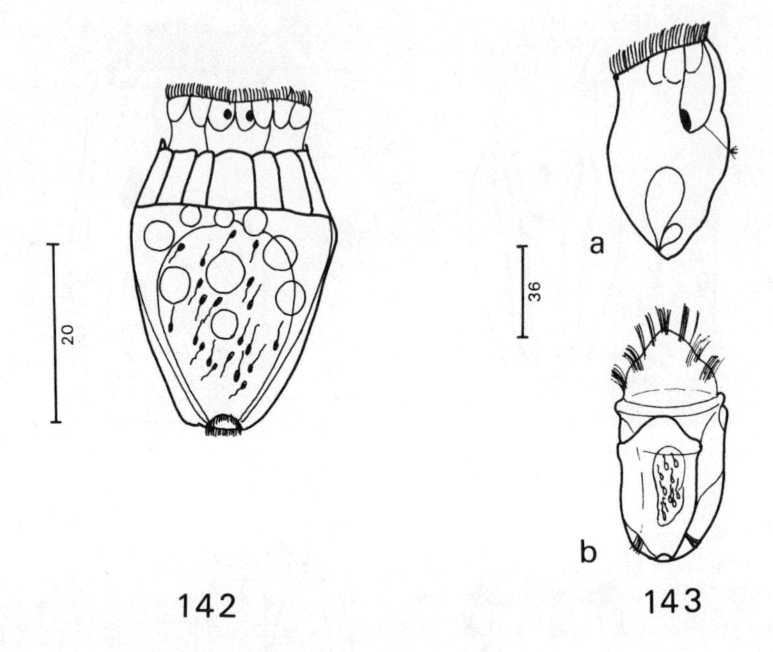

Fig. 142. *Pompholyx sulcata.* Male: dorsal. (After Wesenberg-Lund 1923).

Fig. 143. *Ascomorpha.* Male: *a, A. ecaudis,* lateral; *b, A. ovalis,* dorsal. (*a,* after Wesenberg-Lund 1923, *b,* after Sudzuki 1956).

18(b) Body globular, perhaps without lorica, without furrows of female (fig. 142; female fig. 110). Two red eyes on brain; corona a single row of cilia. No gut; testis very large, pear-shaped; oil globules in body— POMPHOLYX

P. sulcata: length <40 μm; Aug. (fig. 142). *P. complanata:* males unknown. Female, p. 128.

18(c) Body longish, sack-shaped. May have large eye or eyes. Numerous ciliary bunches on head. Cilia on eversible penis (fig. 143). Length <90 μm— ASCOMORPHA

Female, p. 96.

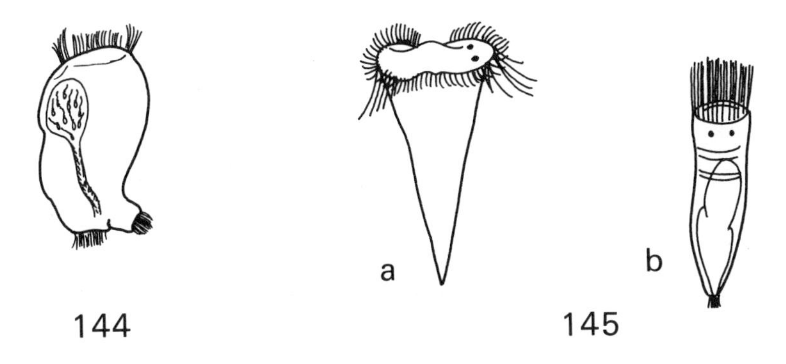

Fig. 144. *Polyarthra vulgaris*. Male: lateral. (After Sudzuki 1955b).

Fig. 145. *Collotheca*. Male: *a*, *C. mutabilis*; *b*, *C. pelagica*. (*a*, after Voigt 1957, *b*, after Rousselet 1893).

19(17) Body globular, posteriorly acuminate, variable in form (fig. 144). Without blades of female (fig. 94). No gut. May have no eyes. Large testis; 2 prostate glands; end of body drawn out to form copulatory organ; oil globules round testis; cilia at body end. Very fast swimmers— POLYARTHRA

Female, p. 109.

— Body longish, spindle-shaped (fig. 145). Body end ciliated; no eversible penis— COLLOTHECA

Males of *C. libera* unknown. Female, p. 142.

For descriptions of males of various genera see Wesenberg-Lund (1923); also Hermes (1932) and Sudzuki (1955b, 1956, 1958). For some information on various aspects of sexuality and dimorphism see Buchner, Mutschler & Kiechle (1967), Buchner, Tiefenbacher, Kling & Preissler (1970), Gilbert (1963, 1968, 1973), Gilbert & Thompson (1968), Kiechle & Buchner 1966.

ACKNOWLEDGEMENTS

I am pleased to record here my gratitude to all those people who have generously given help in the preparation of this key and who have contributed in no small measure to its successful completion.

I am greatly indebted to Mr A. L. Galliford and Mr E. D. Hollowday, who most generously offered me, during the preparation of the key, the benefits of their long experience and knowledge of British rotifers. They have given me advice, criticism, encouragement and information and each has contributed an impressive record list, without which the British list would be much the poorer. To them both I offer my gratitude and thanks.

I am most grateful to Dr Barbara Gilchrist and to Professor J. Green, who each very kindly read the initial version of the key and gave very helpful criticism and encouragement. Dr Gilchrist and her students tried out the early version of the key during classes, and Dr Nan Duncan also made the key available to her students for trial. I thank all those who helped in this way. I also thank Dr Mary Burgis for early encouragement.

Dr Agnes Ruttner-Kolisko generously gave time to read and comment upon the manuscript in the later stages of preparation, and offered valuable criticism and advice, for which I thank her. I am also indebted to Dr J. M. Elliott for his critical reading of the key and helpful suggestions which contributed to many improvements. Mr J. E. M. Horne also spent much time reading and editing the manuscript and I am most grateful to him, not only for his advice and his attention to detail, but also for his patience and kindness throughout the preparation of this key. Miss Pam Moorhouse compiled the index.

Record lists were very kindly contributed by Professor Green, Dr Duncan, Dr Kieran Horkan and Dr Margaret Doohan (Mrs P. Frayn). I am greatly indebted to Professor P. Butler for making a microscope available for my use.

Finally I should like to thank my constant companions of collecting trips, Dr A. John Pontin, Katherine, Carol, and Mark Pontin, for their enduring interest and help.

AUTHOR'S ADDRESS:

26 Hermitage Woods Crescent,
St. Johns,
Woking,
Surrey GU21 1UE.

REFERENCES

Ahlstrøm, E. (1938). Plankton Rotatoria from North Carolina. *J. Elisha Mitchell scient. Soc.* **54,** 88-110.

Ahlstrøm, E. (1940). A revision of the rotatorian genera *Brachionus* and *Platyias* with description of one new species and two new varieties. *Bull. Am. Mus. nat. Hist.* **77,** 143-184.

Ahlstrøm, E. (1943). A revision of the rotatorian genus *Keratella,* with descriptions of three new species and five new varieties. *Bull. Am. Mus. nat. Hist.* **80,** 411-457.

Albertova, O. (1959). Variabilität der Borstenlangen bei *Tetramastix opoliensis* Zach. *Věst. čsl. zeměd. Mus.* **23,** 376-380.

Albertova, O. (1960). Eine neue Art der Gattung *Polyarthra. Věst. čsl. Spol. zool.* **24,** 16-18.

Amrèn, H. (1964). Ecological and taxonomical studies on zooplankton from Spitsbergen. *Zool. Bidr. Upps.* **36,** 209-276.

Bartoš, E. (1948). On the Bohemian species of the genus *Pedalia* Barrois. *Hydrobiologia,* **1,** 63-77.

Bartoš, E. (1951a). The key to the determination of Rotifera from the genus *Polyarthra* Ehrenbg. *Čas. národ. Mus.* **6,** 82-91.

Bartoš, E. (1951b). Czechoslovakian species of genera *Testudinella* and *Pompholyx. Sb. Klubu přír. Brně,* **29,** 10-20.

Bartoš, E. (1959). Vířníci — Rotatoria. *Fauna ČSR,* **15,** 1-965. Praha.

Beadle, L. C. (1963). Anaerobic life in a tropical crater lake. *Nature, Lond.* **200,** 1223-1224.

Beauchamp, P. de (1909). Recherches sur les Rotifères. *Archs Zool. exp. gén.* **10,** 4, 1-410.

Beauchamp, P. de (1912). Sur deux formes inférieures d'Asplanchnides. *Bull. Soc. zool. Fr.* **36,** 223-233.

Beauchamp, P. de (1932). Contribution a l'étude du genre *Ascomorpha* et des processus digestifs chez les Rotifères. *Bull. Soc. zool. Fr.* **57,** 428-449.

Beauchamp, P. de (1951). Sur la variabilité specifique dans le genre *Asplanchna. Bull. biol. Fr. Belg.* **85,** 137-175.

Bērziņš, B. (1951). On the collothecacean Rotatoria. *Ark. Zool.* Ser. 2, **1,** 565-592.

Bērziņš, B. (1954a). Nomenklatorische Bemerkungen an einigen planktischen Rotatorien-Arten der Gattung *Keratella. Hydrobiologia,* **6,** 321-327.

Bērziņš, B. (1954b). A new rotifer, *Keratella canadensis. J. Quekett microsc. Club,* **4,** 113-115.

Bērziņš, B. (1955). Taxonomie und Verbreitung von *Keratella valga* und verwandten Formen. *Ark. Zool.* Ser. 2, **8,** 549-559.

Bērziņš, B. (1960). Rotatoria I. Genus: *Synchaeta,* and II. Genus: *Trichocerca.* Zooplankton, Sheet 84, 1-7 and 85, 1-3. Cons. int. Explor. Mer.

Bērziņš, B. (1962). Revision der Gattung *Anuraeopsis* Lauterborn. *K. fysiogr. Sällsk. Lund Förh.* **32,** 33-47.

Birky, C. W. & Gilbert, J. J. (1971). Parthenogenesis in rotifers: the control of sexual and asexual reproduction. *Am. Zool.* **11,** 245-266.

Bory de St. Vincent, J.- B. G. M. M. (1822). *Dictionnaire classique d'histoire naturelle.* Vol. 2 (As-Cac), 621 pp. Paris.

Bory de St. Vincent, J.- B. G. M. M. (1824a). *Encyclopédie méthodique. Histoire naturelle des zoophytes, ou animaux rayonnés, faisant suite à l'histoire naturelle des vers de Bruguière.* Vol. 2, 819 pp. Paris.

Bory de St. Vincent, J.- B. G. M. M. (1824b). *Dictionnaire classique d'histoire naturelle.* Vol. 6 (E-Fouq), 593 pp. Paris.

Bory de St. Vincent, J.- B. G. M. M. (1826). *Essai d'une classification des animaux microscopiques.* 104 pp. Paris.

Buchholz, H. & Rühmann, D. (1956). *Notholca* und *Kellicottia. Mikrokosmos,* **45,** 267-270.

Buchner, H., Mutschler, C. & Kiechle, H. (1967). Die Determination der Männchen- und Dauereiproduktion bei *Asplanchna sieboldi. Biol. Zbl.* **86,** 599-621.

Buchner, H., Tiefenbacher, L., Kling, P. & Preissler, K. (1970). Über die physiologische Bedeutung des Magen-Darm-Rudiments der Männchen von *Asplanchna sieboldi. Z. vergl. Physiol.* **67,** 453-454.

Bülow, T. (1954). Ernahrungsbiologische Untersuchungen an Euchlaniden. *Zool. Anz.* **153,** 126-134.

Carlin, B. (1939). Die Rotatorien einiger Seen bei Aneboda. *Meddn Lunds Univ. limnol. Instn,* **2,** 1-68.

Carlin, B. (1943). Die Planktonrotatorien des Motalaström. *Meddn Lunds Univ. limnol. Instn,* **5,** 1-255.

Carlin, B. (1945). Vara Planktonrotatorien. *Medlemsblad för Biologilärarnas Förening,* **1,** 17-23.

Collin, A., Dieffenbach, H., Sachse, R. & Voigt, M. (1912). Rotatoria und Gastrotricha. *Süsswasserfauna Deutschlands,* H.14, 273 pp. Jena. Fischer.

Dieffenbach, H. & Sachse, R. (1912). Biologische Untersuchungen an Rädertieren in Teichgewässern. *Int. Revue ges. Hydrobiol. Hydrogr.* Suppl. ser. 3, **2,** 94 pp.

Dobie, W. M. (1849). Description of two new species of *Floscularia,* with remarks. *Floscularia campanulata* et *F. cornuta. Ann. Mag. nat. Hist.* ser. 2, **4,** 233-238.

Donner, J. (1949). *Horaella brehmi* nov. gen., nov. spec. *Hydrobiologia,* **2,** 134-140.

Donner, J. (1950). Rädertiere der Gattung *Cephalodella. Arch. Hydrobiol.* **42,** 304-328.

Donner, J. (1953). *Trichocerca Ruttneri,* n. spec. *Öst. zool. Z.* **4,** 9-22.

Donner, J. (1954). Zur Rotatorienfauna Südmährens. *Öst. zool. Z.* **5,** 30-117.

Donner, J. (1959). Bemerkungen zur Rädertierart *Synchaeta oblonga* Ehrenb. 1832. *Verh. zool.-bot. Ges. Wien,* **98/99,** 26-30.

Donner, J. (1965). Ordung Bdelloidea. *Bestimm. Büch. Bodenfauna Europas,* Lief. **6,** 1-297.

Donner, J. (1966). Rotifers. [translation by H. G. S. Wright.] 80 pp. London, Warne.

Edmondson, W. T. (1959). Rotifera. In *Freshwater Biology* (H. B. Ward & G. C. Whipple) 2nd edition ed. W. T. Edmondson. 420-494. New York. Wiley.

Ehrenberg, C. G. (1832). Ueber die Entwicklung und Lebensdauer der Infusionsthiere, nebst ferneren Beiträgen zu einer Vergleichung ihrer organischen Systeme. *Abh. Akad. Wiss. Berlin,* 1831, 1-154.

Ehrenberg, C. G. (1834). Dritter Beitrag zur Erkenntniss grosser Organisation in der Richtung des kleinsten Raumes. *Abh. Akad. Wiss. Berlin,* 1833, 145-336.

Ehrenberg, C. G. (1838). *Die Infusionsthierchen als volkommene Organismen. Ein Blick in das tiefere organische Leben der Natur.* Leipzig. 547 pp.

Ehrenberg, C. G. (1853). Ueber die neuerlich bei Berlin vorgekommenen neuen Formen des mikroskopischen Lebens. *Mber. Akad. Wiss. Berlin,* 183-194.

Fergg, I. (1964). Untersuchungen über die Variabilität der Rädertiere 4. Vergleichende biometrische Untersuchungen über die Variation von *Keratella cochlearis* und *K. quadrata. Zool. Anz. Suppl.* **27,** 253-268.

Focke, E. (1961). Die Rotatoriengattung *Notholca* und ihr Verhalten im Salzwasser. *Kieler Meeresforsch.* **17,** 190-205.

Förster, K. A. (1951). *Keratella cochlearis* und *K. quadrata* im jahrlichen Formenwechsel. *Mikrokosmos,* **41** (2) 30-33.

Galliford, A. L. (1950). Rotifera of Lancashire and Cheshire. Report No. 2, 1948-9. *Rep. Lancs. Chesh. Fauna Comm.* No. 29, 108-114.

Galliford, A. L. (1953). Notes on the distribution and ecology of the Rotifera and Cladocera of N. Wales. *NWest Nat.* **24,** 513-529.

Galliford, A. L. (1954a). Rotifera of Lancashire and Cheshire. Report No. 3, 1950-3. *Rep. Lancs. Chesh. Fauna Comm.* No. 30, 69-78.

Galliford, A. L. (1954b). Notes on the freshwater organisms of Lundy with special reference to the Crustacea and Rotifera. *Rep. Lundy Fld Soc.* **7** (1953), 29-35.

Galliford, A. L. (1960). Notes on the microscopic fauna and flora of the Leeds-Liverpool Canal in south-west Lancashire. *Proc. Lpool Nat. Fld Club,* 1959, 23-28.

Galliford, A. L. (1967). Notes on the fauna of Dozmary Pool, Bodmin Moor, Cornwall. *J. Quekett microsc. Club,* **30,** 277-280.

Galliford, A. L. (1971). Further notes on the microfauna and flora of Knowsley Park. *Proc. Lpool Nat. Fld Club,* 1964-9, 4 pp.

Galliford, A. L. (1974). Notes on the Rotifera and Crustacea of the canals, reservoirs, ponds and the River Tame in the Audenshaw and adjacent districts of Manchester. *Publs. Lancs. Chesh. Fauna Soc.* No. 65, 17-22.

Gilbert, J. J. (1963). Contact chemoreception, mating behaviour and sexual isolation in the rotifer genus *Brachionus. J. exp. Biol.* **40,** 625-641.

Gilbert J. J. (1968). Dietary control of sexuality in the rotifer *Asplanchna brightwelli* Gosse. *Physiol. Zool.* **41,** 14-43.

Gilbert, J. J. (1973). The adaptive significance of polymorphism in the rotifer *Asplanchna.* Humps in males and females. *Oecologia,* **13,** 135-146.

Gilbert, J. J. & Thompson, G. A. (1968). Alpha tocopherol control of sexuality and polymorphism in the rotifer *Asplanchna. Science, N.Y.* **159,** 734-738.

Gillard, A. (1947). Het geslacht *Testudinella* in Belgie. *Natuurw. Tijdschr.* **29,** 153-158.

Gillard, A. (1948). De Brachionidae van Belgie. *Natuurw. Tijdschr.* **30,** 159-218.

Gosse, P. H. (1850). Description of *Asplanchna priodonta,* an animal of the class Rotifera. *Ann. Mag. nat. Hist.* ser. 2, **6,** 18-24.

Gosse, P. H. (1851). A catalogue of Rotifera found in Britain, with descriptions of five new genera and thirty-two new species. *Ann. Mag. nat. Hist.* ser. 2, **8,** 197-203.

Gosse, P. H. (1857). Twenty-four new species of Rotifera. *Jl R. microsc. Soc.* 1857, 1-7.

Guerne, J. de (1888). *Excursions zoologiques dans les îles de Fayal et de San Miguel (Açores).* Gauthier-Villars. 112 pp.

Halbach, U. (1970). Die Ursachen der Temporalvariation von *Brachionus calyciflorus* Pallas. *Oecologia,* **4,** 262-318.

Hanley, J. (1949). The narcotisation and mounting of Rotifera. *Microscope,* **7,** 154-159.

Hanley, J. (1954). Permanent preparations of Rotifera. *J. Quekett microsc. Club,* ser. 4, **4,** 18-25.

Harring, H. K. (1913a). A list of the Rotatoria of Washington and vicinity. *Proc. U.S. natn. Mus.* **46,** 387-405.

Harring, H. K. (1913b). Synopsis of the Rotatoria. *Bull. U.S. natn. Mus.* **81,** 1-226.

Harring, H. K. & Myers, F. J. (1924). The rotifer fauna of Wisconsin II. A revision of the notommatid rotifers. *Trans. Wis. Acad. Sci. Arts Lett.* **21,** 415-549.

Harring, H. K. & Myers, F. J. (1926). The rotifer fauna of Wisconsin III. A revision of the genera *Lecane* and *Monostyla. Trans. Wis. Acad. Sci. Arts Lett.* **22,** 315-423.

Harring, H. K. & Myers, F. J. (1930). The rotifer fauna of Wisconsin V. *Euchlanis* and *Monommata. Trans. Wis. Acad. Sci. Arts Lett.* **25,** 353-413.

Hauer, J. (1924). Zur Kenntnis des Rotatoriengenus *Colurella* Bory de St. Vincent. *Zool. Anz.* **59,** 177-189.

Hauer, J. (1929). Zur Kenntnis der Rotatoriengenera *Lecane* und *Monostyla. Zool. Anz.* **83,** 143-164.

Hauer, J. (1938). Zur Rotatorienfauna Deutschlands (VII). *Zool. Anz.* **123,** 213-219.

Hermes, G. (1932). Die Männchen von *Hydatina senta* und *Rhinops vitrea. Z. wiss. Zool.* **141,** 581-725.

Herrick, C. L. (1885). Notes on American rotifers. *Bull. scient. Labs. Denison Univ.* **1,** 43-62.

Hlava, S. (1904). Einige Bemerkungen über die Exkretionsorgane der Rädertierfamilie Melicertidae und die Aufstellung eines neuen Genus *Conochiloides. Zool. Anz.* **27,** 247-253.

Hlava, S. (1905). Beiträge zur Kenntnis der Rädertiere über die Anatomie von *Conochiloides natans. Z. wiss. Zool.* **80,** 282-326.

Hollowday, E. D. (1945-50). Introduction to the study of the Rotifera, parts 1-17. *Microscope,* **5, 6, 7.**

Hudson, C. T. (1871). On a new rotifer. *Mon. microsc. J.* **6,** 121-124.

Hudson, C. T. & Gosse, P. H. (1886). *The Rotifera; or wheel-animalcules.* London. Longmans, Green & Co. 2 vols. 128 and 144 pp. +30 pl.

Hudson, C. T. & Gosse, P. H. (1889). *The Rotifera: or wheel-animalcules. Supplement.* London. Longmans, Green & Co. 64 pp. +4 pl.

Hutchinson, G. E. (1964) On *Filinia terminalis* (Plate) and *F. pejleri* sp. n. (Rotatoria: family Testudinellidae). *Postilla,* **81,** 8 pp.

Hutchinson, G. E. (1967). *A treatise on limnology,* Vol. 2. New York. Wiley. 1115 pp.

Hyman, L. H. (1951). *The Invertebrates: Acanthocephala, Aschelminthes and Entroprocta.* Vol. 3. New York. McGraw-Hill. 572 pp.

Illies, J. (Ed.). (1967). *Limnofauna Europaea.* Stuttgart. Fischer. 474 pp.

Imhof, O. E. (1888). Fauna der Süsswasserbecken. *Zool. Anz.* **11,** 166-172.

Imhof, O. E. (1891). Ueber die pelagische Fauna einiger Seen des Schwarzwaldes. *Zool. Anz.* **14,** 33-38.

Jennings, H. S. (1903). Rotatoria of the United States. II. A monograph of the Rattulidae. *Bull. U.S. Fish Commn,* **22,** 273-352.

Kellicott, D. S. (1879). A new rotifer. *Am. J. Microsc. popular Sci.,* **4,** 19-20.

Kiechle, H. & Buchner, H. (1966). Untersuchungen über die Variabilität der Rädertiere V. Dimorphismus und Bisexualität bei *Asplanchna. Revue suisse Zool.* **73,** 238-300.

Klement, V. (1955). Über eine Missbildung bei dem Rädertier *Keratella cochlearis* und neue Form von *Keratella quadrata. Zool. Anz.* **155,** 321-324.

Lamarck, J. B. P. A. de M. de (1801). *Systéme des animaux sans vertèbres, ou tableau général des classes, des ordres et des genres de ces animaux.* Paris. 432 pp.

Lauterborn, R. (1900). Der Formenkreis von *Anuraea cochlearis* — ein Beitrag zur Kenntnis der Variabilität bei Rotatorien. I. Morphologische Gliederung des Formenkreises. *Verh. naturh. -med. Ver. Heidelb.* n.ser. **6,** 412-448.

Lauterborn, R. (1908). Gallerthüllen bei loricaten Plankton-Rotatorien. *Zool. Anz.* **33,** 580-584.

Leissling, R. (1924). Zur Kenntnis von *Pompholyx sulcata. Zool. Anz.* **59,** 88-110.

Levander, K. M. (1894). Materialen zur Kenntniss der Wasserfauna in der Umgebung von Helsingfors, mit besonderer Berucksichtigung der Meeresfauna. II. Rotatorien. *Acta Soc. Fauna Flora fenn.* **12,** 3, 1-72.

Löffler, H. (1954). Über eine Varietät von *Pedalia fennica* aus Nordwestpersien. *Zool. Anz.* **152,** 144-146.

Mitchell, C. W. (1913). Experimentally induced transition in the morphological characters of *Asplanchna amphora* Hudson, together with remarks on sexual reproduction. *J. exp. Zool.* **15,** 91-127.

Müller, O. F. (1773). *Vermium terrestrium et fluviatilium, seu animalium infusoriorum, helminthicorum et testaceorum, non marinorum, succincta historia.* Havniae et Lipsiae. Vol. 1, pt. 1.

Müller, O. F. (1776). *Zoologiae Danicae prodromus, seu animalium Daniae et Norvegiae indigenarum characteres, nomina, et synonyma imprimis popularium.* Havniae. 282 pp.

Müller, O. F. (1786). *Animalcula Infusoria fluviatilia et marina, quae detexit, systematice descripsit et ad vivum delineari curavit* O. F. Müller. Havniae. 367 pp.

Myers, F. J. (1934a). New Testudinellidae of the genus *Testudinella* and a new species of Brachionidae of the genus *Trichotria. Am. Mus. Novit.* No. 761.

Myers, F. J. (1934b). A new species of Synchaetidae and new species of Asplanchnidae, Trichocercidae and Brachionidae. *Am. Mus. Novit.* No. 700.

Nachtwey, R. (1925). Untersuchungen über Keimbahn, Organogenese und Anatomie von *Asplanchana priodonta. Z. wiss. Zool.* **126,** 239-492.

Nipkow, R. (1952). *Polyarthra* Ehrenbg. im Zürichsee und anderen Schweizer Seen. *Schweiz. Z. Hydrol.* **14,** 135-181.

Nipkow, R. (1961). Die Rädertiere im Plankton des Zürichsees und ihre Entwicklungsphasen. *Schweiz. Z. Hydrol.* **22,** 398-461.

Nitzsch, C. L. (1827). Cercaria. *Allg. Encycl. Wiss. Künste (Ersch u. Gruber), Leipzig,* sect. 1, 16, 66-69.

Olofsson, O. (1917). Süsswasser-Entomostraken und Rotatorien von der Murmanküste und aus dem nördlichen Norwegen. *Zool. Bidr. Upps.* **5,** 259-294.

Pallas, P. S. (1766). *Elenchus zoophytorum sistens generum adumbrationes generaliores et specierum cognitarum succinctas descriptiones cum selectis auctorum synonymis.* Hagae Comitum. 451 pp.

Parise, A. (1961). Sur les genres *Keratella, Synchaeta, Polyarthra* et *Filinia* d'un lac italien. *Hydrobiologia,* **18,** 121-135.

Parise, A. (1963). Osservazioni sull'eterogonia sperimentale in *Euchlanis. Atti Accad. naz. Lincei Rc.,* Cl. Sci., Ser. 8, **35,** 609-615.

Parise, A. (1966). The genus *Euchlanis* (Rotatoria) in the marsh of Fucecchio (Central Italy) with description of a new species. *Hydrobiologia,* **27,** 328-337.

Pejler, B. (1957). On variation and evolution in planctonic Rotatoria. *Zool. Bidr. Upps.* **32,** 1-66.

Pejler, B. (1962). Morphological studies on the genera *Notholca, Kellicottia* and *Keratella*. *Zool. Bidr. Upps.* **33**, 327-422.

Perty, M. (1850). Neue Räderthiere der Schweiz. *Mitt. naturf. Ges. Bern*, 1850, 17-22.

Pontin, R. M. (1964). A comparative account of the protonephridia of *Asplanchna* (Rotifera) with special reference to the flame bulbs. *Proc. zool. Soc. Lond.* **142**, 511-525.

Pontin, R. M. (1966). The osmoregulatory function of the vibratile flames and the contractile vesicle of *Asplanchna*. *Comp. Biochem. Physiol.* **17**, 1111-1126.

Pourriot, R. (1965). Recherches sur l'écologie des Rotifères. *Vie Milieu*, Suppl. **21**, 1-224.

Pourriot, R. (1970). Quelques *Trichocerca* (Rotifères) et leurs régimes alimentaires. *Annls Hydrobiol.* **1**, 155-171.

Rahm, N. (1956). Sur la présence de *Trochosphaera solstitialis* Thorpe en Afrique. *Bull. Inst. fr. Afr. noire*, **18**, A.

Remane, A. (1933). Zur Organisation der Gattung *Pompholyx*. *Zool. Anz.* **103**, 188-193.

Rousselet, C. F. (1889). Note on *Brachionus quadratus*, a new rotifer. *J. Quekett microsc. Club*, ser. 2, **4**, 32-33.

Rousselet, C. F. (1893). On *Floscularia pelagica* sp.n., and notes on several other rotifers. *Jl R. microsc. Soc.* 1893, 444-449.

Rousselet, C. F. (1897). *Brachionus bakeri* and its varieties. *J. Quekett microsc. Club*, ser, 2, **6**, 328-332.

Rousselet, C. F. (1902). The genus *Synchaeta:* a monographic study, with description of five new species. *Jl R. microsc. Soc.* 1902, 269-290, 393-411.

Rousselet, C. F. (1907). On *Brachionus sericus* n.sp., a new variety of *Brachionus quadratus*, and remarks on *Brachionus rubens*, of Ehrenberg. *J. Quekett microsc. Club*, ser. 2, **10**, 147-154.

Rousselet, C. F. (1908). Note on the Rotatorian fauna of Boston, with description of *Notholca bostoniensis* s.n. *J. Quekett microsc. Club*, ser. 2, **10**, 335-340.

Rühmann, D. (1954). Das Planktonrädertier *Keratella*. *Mikrokosmos*, **43**, 266-270.

Rühmann, D. (1965). Die Rädertiergattung *Mytilina*. *Mikrokosmos*, **54**, 173-175.

Ruttner-Kolisko, A. (1938). Die Nahrungsaufnahme bei *Anapus testudo*. *Int. Revue ges. Hydrobiol. Hydrogr.* **37**, 296-305.

Ruttner-Kolisko, A. (1939). Über *Conochilus unicornis* und seine Koloneibildung. *Int. Revue ges. Hydrobiol. Hydrogr.* **39**, 78-98.

Ruttner-Kolisko, A. (1949). Zum Formwechsel- und Artproblem von *Anurea aculeata (Keratella quadrata)*. *Hydrobiologia*, **1**, 425-468.

Ruttner-Kolisko, A. (1959). *Polyarthra* aus den Kapruner Stauseen. *Anz. Osterr. Akad. Wiss. Jg.* **59**, (1) 1-6.

Ruttner-Kolisko, A. (1969). Kreuzungsexperimente zwischen *Brachionus urceolaris* und *Brachionus quadridentatus*. *Arch. Hydrobiol.* **65**, 397-412.

Ruttner-Kolisko, A. (1970). *Synchaeta calva* nov. spec., a new rotifer from the English Lake District. *Int. Revue ges. Hydrobiol. Hydrogr.* **55**, 387-390.

Ruttner-Kolisko, A. (1974). Plankton rotifers: biology and taxonomy [translated from the German]. *Binnengewässer,* **26**, 1 (Suppl.), 146 pp.

Rylov, W. M. (1935). Das Zooplankton der Binnengewässer. *Binnengewässer,* **15**, 1-272.

Schmarda, L. K. (1854). Zur Naturgeschichte Aegyptens. *Denkschr. Akad. Wiss. Wien. Math-Nat. Kl.* **7**, 2, 1-28.

Schrank, F. von P. (1793). Mikroskopische Wahrnehmungen. *Naturforscher, Halle,* **27**, 25-37.

Schrank, F. von P. (1803). *Fauna Boica. Durchgedachte Geschichte der in Baiern einheimischen und zahmen Thiere,* Bd **3**, 2. 372 pp. Nurnberg.

Scopoli, G. A. (1777). *Introductio ad historiam naturalem, sistens genera lapidum, plantarum et animalium hactenus detecta, caracteribus essentialibus donata, in tribus divisa, subinde ad leges naturae.* Pragae. 506 pp.

Seehaus, W. (1930). Zur Morphologie der Rädertiergattung *Testudinella. Z. wiss. Zool.* **137**, 175-273.

Seligo, A. (1900). *Untersuchungen in den Stuhmer Seen.* Danzig. 88 pp.

Semper, C. (1872). Zoologische Aphorismen. *Z. wiss. Zool.* **22**, 305-322.

Sládeček, V. (1955). A note on the occurence of *Hexarthra fennica* Levander in Czechoslovakia. *Hydrobiologia,* **7**, 64-67.

Sládeček, V. (1968). *Collotheca undulata* n.spec. *Zool. Anz.* **182**, 417-420.

Smirnov, N. S. (1928). *Fadeewella* n.sp. eine neue Rotatoriengattung aus dem Ussuri Gebiet. *Zool. Anz.* **79**, (5/6) 129-133.

Stossberg, K. (1932). Zur Morphologie der Rädertiergattungen *Euchlanis, Brachionus* und *Rhinoglena. Z. wiss. Zool.* **142**, 313-324.

Sudzuki, M. (1955a). Studies on the egg-carrying types in Rotifera I. Genus *Pompholyx. Zool. Mag., Tokyo,* **64**, 219-224.

Sudzuki, M. (1955b). On the general structure and the seasonal occurrence of the males in some Japanese Rotifers, I, II, III. *Zool. Mag., Tokyo,* **64**, 126-129, 130-136, 189-193.

Sudzuki, M. (1956). On the general structure and seasonal occurrence of the males in some Japanese Rotifers, IV, V, VI. *Zool. Mag., Tokyo,* **65**, 1-6, 329-334, 415-421.

Sudzuki, M. (1957a). Studies on the egg-carrying types in Rotifera II. Genera *Brachionus* and *Keratella. Zool. Mag., Tokyo,* **66**, 11-20.

Sudzuki, M. (1957b). Studies on the egg-carrying types in Rotifera III. Genus *Anuraeopsis. Zool. Mag., Tokyo,* **66**, 407-415.

Sudzuki, M. (1958). On the general structure and seasonal occurrence of the males in some Japanese rotifers VII. *Zool. Mag., Tokyo,* **67**, 348-354.

Valkanov, A. (1936). Beitrag zur Anatomie und Morphologie der Rotatoriengattung *Trochosphaera* Semper. *Trud. bŭlg. prir. Druzh.* **17**, 177-195.

Varga, L. (1929). *Rhinops fertöensis,* ein neues Rädertier aus dem Fertö (Neusiedlersee). *Zool. Anz.* **80**, 236-253.

Varga, L. (1930). Beitrage zur Biologie von *Rhinops fertöensis*. *Állat. Közl.* **27,** 17-35.

Varga, L. (1933). *Squatinella geleii* n.sp., ein neuer Rädertier aus Ungarn. *Allat. Közl.* **30,** 177-183.

Varga, L. (1936). *Collotheca balatonica,* ein neues pelagisches Rädertier. *Arb. ung. biol. ForschInst.* **8,** 178-182.

Voigt, M. (1904). Die Rotatorien und Gastrotrichen der Umgebung von Plön. *Plöner Ber.* **11,** 1-166.

Voigt, M. (1957). *Rotatoria. Die Rädertiere Mitteleuropas.* Berlin. Borntraeger. 508 pp.

Wälikangas, I. (1924). Über die Verbreitung von *Pedalion oxyure* Sernow. *Int. Revue ges. Hydrobiol. Hydrogr.* **12,** 339-341.

Waniczek, H. (1930). Untersuchungen über einigen Arten der Gattung *Asplanchna* Gosse. *Annls Mus. Zool. pol.* **8,** 109-322.

Weber, E. F. (1898). Faune Rotatorienne du bassin du Léman. *Revue suisse Zool.* **5,** 263-785.

Weisse, J. F. (1847). Viertes Verzeichniss St Petersburgischer Infusorien, nebst Beschreibung zweier neuer Arten. *Bull. Cl. phys.-math. Acad. imp. Sci. St Petersburg,* **6,** cols. 110-112.

Wesenberg-Lund, C. (1923). Contributions to the biology of the Rotifera. I. The males of the Rotifera. *K. danske Vidensk. Selsk. Skr.* (nat. math. Afd.) 8, **4,** 3, 190-345.

Wesenberg-Lund, C. (1930). Contributions to the biology of the Rotifera. II. The periodicity and sexual periods. *K. danske Vidensk. Selsk. Skr.* (nat. math. Afd.) 9, **2,** 1, 1-230.

Wierzejski, A. (1892). Zur Kenntniss der *Asplanchna*-Arten. *Zool. Anz.* **15,** 345-349.

Wiszniewski, J. (1929). Zwei neue Rädertierarten: *Pedalia intermedia* n.sp. und *Paradicranophorus limosus* n.g.n.sp. *Bull. int. Acad. pol. Sci. Lett.* Ser. B. II. 1929, 137-153.

Wulfert, K. (1938). Die Rädertiergattung *Cephalodella. Arch. naturgesch.* **7,** 137-152.

Wulfert, K. (1939). Beiträge zur Kenntnis der Rädertierfauna Deutschlands IV. *Arch. Hydrobiol.* **35,** 563-624.

Wulfert, K. (1965). Revision der Rotatoriengattung *Platyias* Harring 1913. *Limnologica,* **3,** 41-64.

INDEX

Page numbers in **bold** type indicate main references to rotifers included in the key; those in *italic* type indicate illustrations.